Introduction to Biometrical Genetics

Introduction to Biometrical Genetics

KENNETH MATHER

C.B.E., D.Sc., F.R.S.

Professor of Genetics in the University of Birmingham
(Formerly Vice-Chancellor and Professor of Genetics in the University of Southampton)

JOHN L. JINKS

D.Sc., F.Inst. Biol., F.R.S.

Professor and Head of Department of Genetics in the University of Birmingham

CORNELL UNIVERSITY PRESS
ITHACA, NEW YORK

First published 1977 by Cornell University Press

Set by Hope Services, Wantage,
and printed in Great Britain
at the University Printing House, Cambridge

International Standard Book Number 0-8014-1123-8

Library of Congress Catalog Card Number 77-76809

For information address Cornell University Press,
124 Roberts Place, Ithaca, New York 14850.

Contents

Preface

In the second edition of *Biometrical Genetics*, which appeared in 1971, we set out to give a general account of the subject as it had developed up to that time. Such an account necessarily had to be comprehensive and reasonably detailed. Although it could be, and indeed has been, used by those who were making an acquaintance with this branch of genetics for the first time, it went beyond their needs. We have been encouraged therefore to write an introduction to the genetical analysis of continuous variation aimed primarily at senior undergraduate and postgraduate students, and concentrating on basic considerations, basic principles and basic techniques. This has meant, of course, omitting all reference to some phenomena of more restricted interest, notably sex-linkage, maternal effects, haploidy and polyploidy. It has meant, too, that even with some phenomena which have been included, like interactions, linkage and effective factors, the discussions cannot go into full detail. Anyone who is interested, however, can find further information in *Biometrical Genetics*, to which detailed references have been given where it appeared that these would be helpful.

The order of presentation has been changed with the aim of making it easier for beginners. It is now presented basically in terms of phenomena, additive-dominance variation being taken first, followed by genic interaction, correlated gene distributions and genotype × environment interaction, rather than in terms of the type of data to be analysed, with means first followed by second degree statistics. We believe that this will be found to be more acceptable to the student and will enable him to master the basic phenomena in all their manifestations before proceeding to those which add complexities to the fundamental models and analyses. We have, however, continued to defer consideration of populations until after that of crosses between true-breeding lines, since, although historically populations were dealt with by Fisher before simple crosses, the restrictions on the information to be gained from populations and the possibilities as well as the limitations of its interpretation cannot be appreciated until the analysis and interpretation of data from simple crosses are understood. In this, of course, biometrical genetics follows the pattern already set by classical genetics.

We have taken some of the examples we use from the earlier book, but we have sought wherever possible to use new illustrative material. And although our aim has been to simplify the presentation, we have taken the opportunity in a number of places to bring in relevant advances made since *Biometrical Genetics* was written some six years ago. We have assumed that the reader is familiar with basic genetics and basic statistics.

Biometrical genetics is still too widely regarded as an esoteric form of genetical endeavour, tortuous, over-difficult and of little but theoretical interest. Basic misapprehensions still appear to be abroad, such as that it requires the assumptions of normal frequency distributions and simple additivity in action of the genes and the environment if its analyses are to be meaningful. We hope that this book will help to dispel such notions. We hope too that it will assist the student to a balanced appreciation of biometrical genetics, its theoretical structure and its analytical methodology, its aims and its approach, its capabilities and its limitations, and above all its unique value in practical situations that many geneticists, especially applied geneticists, inevitably encounter.

We are indebted to Dr P. D. S. Caligari for his help in the preparation of the script, and to the Leverhulme Trust Fund for financial assistance during the writing of this book.

November 1976

K.M.
J.L.J.

1

The genetical foundation

1. Continuous variation

Mendel laid the foundation of genetics by the study of differences which divided his peas into sharply distinct categories. Thus there was never doubt as to whether one of his plants was tall or short, or its flowers red or white and so on: the categories did not overlap. He was able to show that each phenotypic class corresponded to one, or at any rate only a few, genotypes and that where there was more than one genotype in the phenotypic class they could be separated by further appropriate breeding tests, that is by the clearly distinguishable classes of plant to which they gave rise among their descendants following appropriate test matings. He was thus able to infer the genes, or factors as he called them, upon whose behaviour hereditary transmission depends, and it has been by the further study of such gene differences in many species of plants and animals that our knowledge of the genetic materials has largely been built up. We should note, however, that plants or animals may differ in this sharply distinct way for reasons other than the genes they carry; in fact because of the environments in which they have lived their lives. Thus the water crowfoot, *Ranunculus aquatilis*, has quite different leaves when growing in running water than when growing on land. In such a case, of course, observation of the environments suggests at once that the difference is not genetic, or at least not wholly genetic, in its causation; but in general an appropriate breeding test is necessary to establish this point.

Now, differences by which individuals are divided into sharply distinct categories are not the only variation to be seen in either natural populations or experimental families. Mendel's peas themselves showed further variation, for his talls ranged from 6 to 7 ft or even more in height and his shorts from 9 to 18 inches (see Bateson, 1909). The important thing for his experiments and their interpretation was that despite the variation within the classes, the talls and shorts did not overlap in height: each individual could be classified unambiguously as tall or short. There was in fact a discontinuity in the distribution of heights between tall and short, all plants below the discontinuity being short and all above it tall; and as Mendel showed, they differed correspondingly and consistently in their genotypes.

The same complexity of variation can be seen in other species. For example, in man we can recognize dwarf individuals which owe their character to a single gene difference from normals, from whom they are generally clearly distinguishable in respect of stature. Yet people who are not distinguishable in this way - those of normal stature - are not all alike. Indeed they range widely in stature; but the variation they show is of a different kind, with every stature represented between wide limits. The middle statures of the range are the most common and if we examine a large number of individuals we find that the gradations from one stature to the next are so fine as to be almost imperceptible. There are in fact no discontinuities in the distribution of normal stature: the variation is continuous.

Such continuous variation is ubiquitous in living things and, apart perhaps from a few special cases like antigenic specificity, it is displayed by all characters. Thus in general there is no distinction between continuous and discontinuous variation in the characters by which they are displayed and indeed, as we have already seen, we quite commonly observe the two kinds of variation side by side in the same family or population. So, whatever the reasons for the differences between the two kinds of variation, they are not mutually exclusive.

Some examples of continuous variation are shown in Fig. 1. In principle the number of classes into which individuals can be divided according to the manifestation of the character is limited only by the accuracy of the measurements we can make. We find it convenient, however, to group the individuals whose measurements fall between certain limits, which we choose for our own convenience, and represent the variation by recording the numbers falling into the various classes defined in this way. We then obtain histograms as illustrated in Figs. 1 (a) and (c) from which the general shape of the distribution resulting from the variation can be seen. It should nevertheless be remembered that the grouping we are using is purely arbitary: it does not spring from discontinuities in the variation itself and so provides no basis for an analysis of the causes of variation in the way that Mendel showed to be possible with discontinuous variation.

One class of character, however, requires a special word. Sometimes the very nature of the character itself imposes certain discontinuities on the variation it shows. Thus the number of vertebrae in a vertebrate

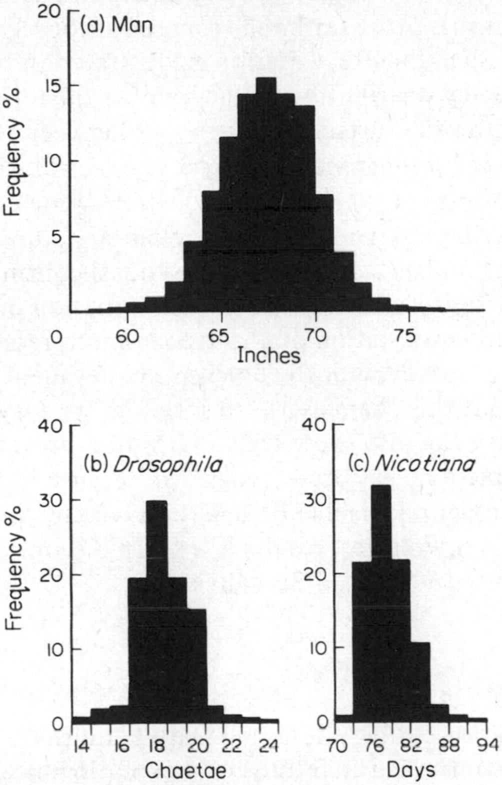

Fig. 1. Frequency distributions illustrating three examples of continuous variation. (a) Stature, in inches, of 8585 men; (b) number of sternopleural chaetae in 200 individuals of *Drosophila melanogaster*; (c) time of first flowering, in days after sowing, in 200 individuals of *Nicotiana rustica.* In all cases the frequencies are expressed as percentages of the total number of individuals observed. In (a) and (c) the discontinuities of the histograms are imposed on the distribution by artificial grouping of the observations, for purposes of representation: the characters are truly continuous in their variation. In (b) however, the discontinuities of the histogram arise from the nature of the character, since we cannot recognize fractional chaetae: the variation is quasi-continuous. The frequencies in (a) and (b) accord with the normal distribution, but (c) departs from the normal in that it shows positive skewness.

animal, or bristles on an insect, can display only a specific set of values, for the number must be an integer since fractional vertebrae or bristles are ruled out. Such a character is said to be meristic and its variation quantal, the expression changing by quanta and not smoothly as in truly continuous, or as it is often (although somewhat loosely) called, quantitative variation. Such quantal variation is illustrated in Fig. 1 (b) which shows the frequency distribution of the number of sternopleural chaetae (bristles situated on the surface of the thorax between the front and midlegs) in a line of *Drosophila melanogaster.* The distribution is very like the histograms of Figs. 1 (a) and (c) although now the class limits are not set arbitrarily by the observer but by the quantal nature of the character itself. Such variation has been described as 'quasi-continuous' (Grüneberg, 1952) because it suggests a truly continuous variation of an underlying potential for the manifestation of chaetae, an interpretation which, as we shall see, accords well with the extensive experimental information that we have about this character in *Drosophila.* We may note too that this very character can also show truly discontinuous variation, for the gene Sp (Sternopleural) has been recognized because it increases the number of sternopleural chaetae to an extent which, at any rate in flies raised at higher temperatures, results in a sharp discontinuity of chaeta number between wild-type and Sp individuals.

2. The genic basis

Continuous variation is ubiquitous and Darwin himself emphasized the significance for continuing adaptative and evolutionary change of the small cumulative steps which it makes possible. It is important too for plant and animal breeders since it is as characteristic a feature of the commercially important characters - yield, fertility, quality, conformation and so on - of domesticated species as it is of the biologically significant characters on which depends the success of a species in the wild. Means of analysing such variation and especially of uncovering the way in which the genetic materials play their part in its determination is thus of prime importance to both our understanding of organisms in the wild and our manipulation of them for practical purposes under domestication. At the same time, the Mendelian approach is denied to us by the absence of those clearly distinct classes from whose contrasts genes can be inferred and from whose frequencies the properties of these genes can be investigated. How then are we to proceed?

Clearly our approach must start with the frequency distribution, to which a continuously varying character gives rise when it is observed in a number of individuals, as illustrated in Fig. 1. Such a distribution is characterized by certain statistics of which its mean and variance are the most important for our purpose, and to which we can add the relevant covariances or correlations where the simultaneous distributions of related individuals are available. If we can find a way of interpreting and understanding these means, variances and covariances in genetical terms they provide the analytical tool that we need.

This approach was pioneered by Galton (1889) in the attempt that he made to elucidate the principles of heredity in the days before genetics as we know it had come to life with the so-called 'rediscovery' of Mendelism in 1900. Galton's investigations were continued and extended by Pearson, and their application of statistical mathematics to biological problems marked a significant step in the growth of that aspect of quantitative biology which we now call biometry or biometrics. They showed us the quantities in terms of which continuous variation can be analysed and Galton was indeed able to demonstrate through the calculation of correlations between relatives (a concept which he introduced) that there must be an hereditary component in continuous variation. He got no further, however, and little progress was in fact made in understanding the genetical implications of these statistical quantities until, in a classical treatise published in 1918, R. A. Fisher showed how the biometrical findings not only could be interpreted, but indeed in some respects virtually demanded interpretation, in terms of Mendel's factors, by then termed genes and known to be carried on the chromosomes. In so bringing together the Galtonian approach and the Mendelian basis, Fisher laid the foundation of what we know as biometrical genetics.

The first great principle of genetics is that the phenotype is the resultant of the individual's genotype and the environment in which that individual develops and lives its life. The phenotype can thus be altered by both change in the genotype and change in the environment. We would thus expect there to be an element in continuous variation that sprang from variation of the environment as well as an element depending on differences among the genotypes. That this was indeed the case was first demonstrated by Johannsen (1909) from his observations on the dwarf bean (*Phaseolus vulgaris*), which shares with many other species of plant, including Mendel's peas, the property of regular self-pollination. Given Mendelian heredity therefore, we would expect individuals generally to be homozygous for their genes. All their progeny would thus be genetically alike, and would constitute what Johannsen called a pure line; although of course different pure lines might be expected to be genetically different in being homozygous for different genes. Johannsen isolated 19 such pure lines and he was able to show that when comparisons were made between the lines, the average weights of daughter beans were related to those of their parents, but that when comparisons were made within lines there was no such relation (see Darlington and Mather, 1949). Thus bean weight, a continuously variable character, showed only non-heritable variation within lines, but there was a genetical component in the differences between lines.

Thus the heritable and non-heritable differences were jointly responsible for the variation in seed weight of the beans; they were of the same order of magnitude in their effects; and they could be distinguished only by a breeding test. All the many analyses of continuous variation undertaken over the years on many characters in many species, both plant and animal, have revealed this combination of heritable and non-heritable agencies in the determination of continuous variation. We return to it later, but one further point remains to be noted now. The distribution of sternopleural chaetae shown in Fig. 1b is from a pure line of *Drosophila melanogaster*, which was produced by inbreeding over many generations. The variation is therefore all non-heritable. Now a fruit fly has sternopleural chaetae on both sides of its thorax and the numbers of chaetae on the two sides, when averaged over many individuals, are alike. Yet in a single individual they are not always exactly the same, differing frequently by one or two chaetae and at times by even more (Table 1). It is difficult to attribute these differences to differences in the external environmental agencies impinging on the two sides of the fly, or rather of the larva from which it developed. The differences are much more plausibly attributable to the vagaries of development, in cell division and so on, affecting the two sides differently. The bilateral difference is thus generally taken as a measure of the stability, or instability if one looks at it the other way, of the developmental processes. They are non-heritable differences but are not due to environmental differences in the strict sense. Furthermore, an analysis of variance of the chaeta numbers of flies from an inbred line shows that the variation between flies, though higher than that between the sides of the same fly (thus revealing the action of environmentally determined differences between individuals), are not markedly higher (Table 1). Thus the non-heritable differences that we can observe between individuals in this or any other species, are not always and not wholly to be attributed to differences in the environment: they may in part, even in large part, be reflecting an instability of development.

Turning now to the heritable component of the variation, it was observed by Galton that not only was there a correlation between parent

TABLE 1.

Non-heritable variation for sternopleural chaeta number between and within females of the Samarkand inbred line of *Drosophila melanogaster*.

Chaeta no. (sum of sides)	17	18	19	20	21	22	23	24	Total
Number of flies	1	11	31	55	55	36	25	7	221

Mean Chaeta no. 20.787

Difference between sides	0	1	2	3	4	Total
Number of flies	61	96	49	13	2	221

Mean difference 1.090

Analysis of variance

	df	MS
Between flies	220	2.196
Within flies (= between sides)	221	1.996

1.996/2.196 = 91% of the variation between flies is a reflection of the developmental variation within flies.

and offspring in their manifestations of the character he was observing (usually some morphological character, like stature in man); but that the correlation was the same between male parents and their offspring as it was between female parents and their offspring. This strongly suggests that both parents contribute equally to the heredity of their offspring as reflected in the variation under observation, in other words that the hereditary element is transmitted equilineariy from the two parents in continuous variation just as it is with Mendelian genes. This equilinearity of transmission has been confirmed time and time again in experiments where reciprocal crosses made between two parents have produced families which, apart from differences attributable to sampling variation, were alike in their mean expressions of the character and in their variances also. Reciprocal differences are seen no more commonly in the study of continuous variation than in any other kind of genetical investigation, and when they do appear it is chiefly where the study of Mendelian genes warns us to expect them to appear, for example where the unequal transmission of sex chromosomes might be expected to be involved.

The equilinearity of relationship between parent and offspring gives a strong presumption that the heritable element of continuous variation reflects the effects of genes transmitted in the same way as Mendelian genes, that is by the chromosomes, but acting in some way to produce this quantitative type of variation. That this is indeed the case has been amply demonstrated by experiments, particularly in *Drosophila melanogaster* where the experimental analysis can be taken further than in other species. In this fly, inversions are available in each of the three major chromosomes (X, II and III) which largely, although in most cases not entirely, suppress recombination of genes between the inverted chromosome and its normal wild-type counterpart in heterozygous females. These inversion chromosomes can also be marked by dominant mutants. The marker genes make it possible to follow the marked chromosomes from one generation to another, and the inversions ensure that the marked chromosomes are transmitted as units largely free from genic erosion by recombination when they are kept heterozygous with their normal counterparts. In consequence these chromosomes are of great use in a variety of ways for analysing genetical differences.

Mather and Harrison (1949) had twelve lines, all wild-type, but ranging from 36.00 to 70.25 in their average numbers of the abdominal chaetae, or sternites, borne on the ventral surfaces of the 4th and 5th abdominal segments - a character which shows quasi-continuous variation like that of the sternopleural chaetae. They crossed each of their lines to a tester stock which carried inversions in all three major chromosomes, each of which was marked by a dominant gene, the X by Bar eye-shape (B), II by Plum eye-colour (Pm) and III by Stubble bristles (Sb). The abdominal chaetae numbers were determined for the F_1 female flies that were heterozygous for the B, Pm and Sb chromosomes. These F_1 chaetae numbers differed, of course, from those of the parent wild-type lines, because unlike the parents they were not homozygous for the wild-type chromosomes but heterozygous for the wild-type and the marked chromosomes. The B, Pm, Sb F_1 females were then back-crossed each to its wild-type parent line. The resulting families contained eight classes of daughters, distinguishable by the segregation of the X chromosome marked by B, II marked by Pm and III marked by Sb. We need, however, note only that, apart from the effects of any recombination the inversions had failed to suppress and from the effects of the small chromosome IV which was not followed in the experiment, the wild-type daughters would be genetically like the original parent line, since they were carrying none of the marked chromosomes, while the B, Pm, Sb progeny would be like the F_1, heterozygous for the marked and wild-type homologues of X, II and III. The chaetae numbers of these classes were also determined.

The differences in chaetae number (*y*-ordinate) between the two classes in the back-cross progenies from the twelve lines are plotted in Fig. 2 against the differences in chaeta number (*x*-abscissa) between the

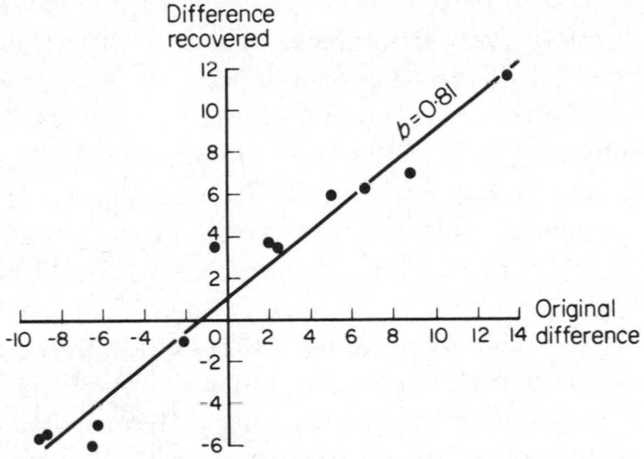

Fig. 2. Mather and Harrison's (1949) data relating the genetical component of variation for the number of abdominal chaetae in *Drosophila melanogaster* to the chromosomes. The slope of the regression shows that 81% of the variation in chaeta number is unambiguously ascribable to genes borne by the three major chromosomes, which on allowing for genes which the experiment could not be expected to pick up accords with all the heritable variation being mediated by nuclear genes.

parent lines and their respective F_1s. A negative value of *x* indicates that the parent line had fewer chaetae than its F_1, and a positive value that it had more. Negative and positive values appeared of course when wild-type lines with low and high chaetae numbers respectively were compared with their F_1s, and the size of the difference reflects the heritable contributions of the wild-type chromosomes since the marked chromosomes were the same in all the crosses. A negative value of *y* indicates a similar shortage of chaetae on the wild-type progeny in the back-cross by comparison with their B, Pm, Sb sisters, and a positive value indicates a corresponding excess. There is a direct relation between *y* and *x*, the regression of *y* on *x* being 0.8073. This means that for every difference of one chaeta between parent and F_1, a difference of 0.8 of a chaeta was recovered in the back-cross.

The implications of these results are clear. First, hereditary elements mediating the continuous variation of abdominal chaeta number must segregate just like Mendelian genes since the differences between the parents and F_1 reappear within the back-cross families. Secondly, since these differences reappear between just those classes whose chromosome constitutions are like those of parents and F_1, the hereditary units in question must be borne in the chromosomes. Cytoplasmic units cannot in any case be involved in the recovered differences as the two types of back-cross fly were from the same mother in each case. Thirdly and finally, since 81% of the parent-F_1 difference was recovered on average, genes carried by these three major chromosomes must be responsible for a minimum of 81% of the heritable differences in abdominal chaetae number among the parent lines. We should however recall that inversions do not always fully suppress recombination, and in these experiments the inversions used in chromosome III would probably suppress recombination in only one of the two arms of the chromosome, while that in the X would allow some recombination in the centre of the chromosome. At the same time we should bear in mind that the small chromosome IV was not controlled, and any difference due to its genes would not be recovered consistently between the wild-type and B, Pm, Sb progeny in the back-cross. The recovery of 81% of the differences actually achieved, therefore, makes it very likely that all the hereditary determinants of the variations in abdominal chaeta number are carried by the chromosomes. In other words the hereditary element in continuous variation springs from genes borne on the chromosomes in just the same way as the genes familiar from Mendelian analysis.

3. Assaying the chromosomes

Marked chromosomes can also be used to build up homozygous lines which carry the three major chromosomes from any two wild-type stocks in all the eight possible combinations. The wild-type stocks are crossed with that carrying the marked chromosomes, and the wild-type chromosomes are carried heterozygous against their marked homologues until they have been brought together in each of the eight combinations. Similar heterozygotes are then mated together and their wild-type progeny, which will be true breeding for the relevant combination of wild-type chromosomes, are used as the foundation of the desired line, the marked chromosomes being thus eliminated at the last stage in the construction of each line. Caligari and Mather (1975) have used this approach in the analysis of the differences in sternopleural chaeta number between two inbred lines, Samarkand (Sam) and Wellington (Well). Denoting the Well chromosomes X, II and III by WWW and those of Sam correspondingly by SSS, the eight homozygous lines WWW, WWS, WSW, WSS, SWW, SWS, SSW and SSS were built up using appropriate marked chromosomes, WWW and SSS being of course reconstructions of the Well and Sam lines from which the chromosomes were originally taken. The extent to which the WWW and SSS lines differ from Well and Sam is a measure of the recombination that went on between the wild-type and the marked chromosomes during the construction of the eight lines and also, of course, of any effect of the small chromosome IV which was again not controlled in the experiment.

As part of a larger experiment Caligari and Mather raised these eight lines, together with Well and Sam, in three types of culture container at a temperature of 21.5°C. All cultures were replicated so as to yield an estimate of error variation. The three types of culture container differed a little in the mean numbers of chaetae borne by the flies they yielded, but there was no evidence that the eight lines reacted differentially to these effects of the containers and the results have therefore been averaged over containers as well as over replicate observations. The means of the eight lines are shown in Table 2. The first point to note is that SSS exceeded WWW by an average of 19.717 - 18.350 = 1.367 chaetae, whereas Sam exceeded Well by 1.908 chaetae. There has thus been a 72% recovery of the parental difference in the reconstituted SSS and WWW lines. Now during the construction of the eight lines every wild-type chromosome was kept heterozygous with its marked homologue for at least four generations and so had at least four opportunities of recombining with it, by contrast with the single opportunity for recombination in the experiment of Mather and Harrison discussed above. So despite the use of marked chromosomes more effective in their suppression of recombination than those of Mather and Harrison, the greater number of opportunities for recombination has resulted in some loss of the parental difference; but again it was a sufficiently small reduction to be consonant with the hereditary element in the variation of sternopleural chaeta number arising from genes borne on the chromosomes.

We can, however, take the analysis further. Since all combinations of the X, II and III chromosomes from Well and Sam are present equally in the eight lines, we can obtain estimates of the effects on chaeta number of the gene differences in each of the three chromosomes, by the use of

TABLE 2.

Sternopleural chaeta numbers in the eight substitution lines from the inbred stocks Samarkand and Wellington of *Drosophila melanogaster* raised at 21·5°C

Difference in chaeta number Sam − Well = 20.650 − 18.742 = 1.908
SSS − WWW = 19.717 − 18.350 = 1.367

Substitution line	Mean chaeta number				
	Observed	Expected 1	O-E1	Expected 2	0-E2
WWW	18.350	18.296 $(m-d_x-d_2-d_3)$	0.054	18.394 $(m-d_x-d_{2W}-d_{3W})$	−0.044
WWS	18.925	19·613 $(m-d_x-d_2+d_3)$	−0.688	18.881 $(m-d_x-d_{2W}+d_{3W})$	0.044
WSW	18·625	17.850 $(m-d_x+d_2-d_3)$	0.775	18.581 $(m-d_x+d_{2W}-d_{3W})$	0.044
WSS	19·025	19.167 $(m-d_x+d_2+d_3)$	−0·142	19.069 $(m-d_x+d_{2W}+d_{3W})$	−0.044
SWW	18.650	18.800 $(m+d_x-d_2-d_3)$	−0.150	18.702 $(m+d_x-d_{2S}-d_{3S})$	−0.052
SWS	20.900	20.117 $(m+d_x-d_2+d_3)$	0.783	20.848 $(m+d_x-d_{2S}+d_{3S})$	0.052
SSW	17.675	18.354 $(m+d_x+d_2-d_3)$	−0.679	17.623 $(m+d_x+d_{2S}-d_{3S})$	0.052
SSS	19.717	19.671 $(m+d_x+d_2+d_3)$	0.046	19.769 $(m+d_x+d_{2S}+d_{3S})$	−0.052
Overall	$m = 18.983$				

Overall $d_x = 0.2521$, $d_2 = -0.2229$, $d_3 = 0.6583$ ± 0.0618

With X chromosome from

	Sam	Well	
	$d_{2S} = -0.5396$	$d_{2W} = 0.0938$	± 0.0874
	$d_{3S} = 1.0729$	$d_{3W} = 0.2438$	

orthogonal functions such as are employed in the analysis of variance (see Mather, 1967). The effect of the X chromosome, for example can be found as $\frac{1}{8}$ (SSS + SSW + SWS + SWW − WSS − WSW − WWS − WWW). Substituting the observed line means from Table 2 we then find

$d_x = \frac{1}{8}(19.717 + 17.675 + 20.900 + 18.650 - 19.025 - 18.625 - 18.925 - 18.350) = 0.2521$ which means that any line carrying the Sam X chromosome will exceed the overall mean of the experiment ($m = 18.9833$) by 0.2521 chaeta because of this chromosome, while any line carrying the Well X will similarly fall short of the overall mean by 0.2521 chaeta. The effects, d_2 and d_3, of chromosomes II and III can be found similarly, using the appropriate functions, and are shown just below the main body of Table 2, together with the relevant standard error based on the estimate of error variance obtained from the replication of the observations referred to above. This standard error applies to the estimates of effect of all three chromosomes, and all three d's are significant. Thus all three chromosomes must be carrying genes influencing the average number of sternopleural chaeta. It will be observed too that d_2 has a negative sign, whereas both d_x and d_3 are positive. Now in finding d_x we gave lines carrying the Sam X a positive sign and those carrying the Well X a negative one. Thus a positive value for d_x means that the Sam X mediated a higher chaeta number than the Well X. Similarly the positive value for d_3 means that the Sam III chromosome gives a higher chaeta number than Well III. The negative value for d_2 means, however, that the Sam II chromosomes give a lower chaeta number than Well II. So the X and III chromosomes in the parental lines are reinforcing each other in their effects on chaeta number, but the II chromosome is acting in the opposing direction.

We can construct expected values for the average chaeta number of each line from the overall mean of the experiment, m, and d_x, d_2 and d_3. Thus the expected value for SSS $= m + d_x + d_2 + d_3 = 18.9833 + 0.2521 + (-0.2229) + 0.6583 = 19.671$. The expectations are shown in column three of Table 2 and the differences between them and the observed means in column 4. There is broad overall agreement with expectation and in some lines, notably SSS and WWW, the observed and expected values of the means agree well, but in other cases, notably WWS, WSW, SWS and SSW, the numerical agreement is not nearly so good. The reason for this is to be seen if we look at the way the chromosomes combine together to produce their effects. Chromosomes II and III show no influence on one another's effects on chaeta number: the difference in effect between the Sam II chromosome and Well II is the same no matter which chromosome III they are with, and vice versa. In other words the effects of these chromosomes simply add on to one another, and we can arrive at the joint effect of either II with either III by summing their individual effects, taking the sign into account, of course.

Thus the deviation from the mean resulting from the combination Well II and Well III is $-d_2-d_3$, that from WS is $-d_2+d_3$, that from SW is d_2-d_3 and that from SS is d_2+d_3. Effects summing in this way are said to be additive, or in the statistician's terminology they show no interaction.

The situation is different, however, when we look at the effects of these two chromosomes in relation to that of the X, for neither II nor III produce as big a difference when present with the Well X as they do when present with the Sam X. The effects of chromosomes II and III in the presence of each of the two X's are shown at the bottom of Table 2, d_{2S} and d_{3S} being the effects of II and III respectively with Sam X, and d_{2W} and d_{3W} their effects with Well X. The value of d_{2S} is, of course, found as 1/4(SSS+SSW−SWS−SWW) = 1/4 (19.717 + 17.675 − 20.900 − 18.650) = − 0.5396 and so on. The effect of chromosome II does not differ significantly from zero when it is with Well X but is quite large with Sam X. This relation is like that termed epistatic in classical genetics (see Fig. 38, Darlington and Mather, 1949). Chromosome III produces an effect even in the presence of Well X but it is only a quarter of that produced in the presence of Sam X. It should be observed, however, that this influence of the X on the effects of II and III does not alter the additive relations holding between II and III themselves: d_{2S} and d_{3S} are additive just as are d_{2W} and d_{3W}. The mean numbers of chaetae expected for the eight lines, allowing for the interactions of X with II and III by the use of d_{2S} etc., are shown in the fifth column of Table 2, and the differences between these and the means observed are given in the sixth column. These differences are now quite small in every case: there is good agreement between observation and expectation. When allowance is made for the ways in which the genes in the different chromosomes interact with one another in producing their effects (interactions which, it should be observed, have their counterparts in classical genetics) the variation in sternopleural chaeta number is accountable in terms of these genes.

4. Locating the genes

The foregoing experiment has led us to recognize that each of the three chromosomes whose effects have been assayed carried one or more genes affecting the number of sternopleural chaetae. Since, however, inversions have been used in the marked chromosomes to suppress recombination as far as possible, each chromosome has behaved as a unit in hereditary transmission and we cannot tell whether its effect was due to only a single gene difference or to more than one, and if to more than one, how these were distributed along the chromosome. In order to take the analysis further we must turn to a different procedure, which is in essence analogous to the three-point experiment by which in classical genetics the locus of a third gene is ascertained in relation to two genes of known loci. Two marker genes, capable of being recognized and followed in transmission by the familiar methods of classical genetics and of known positions in the linkage map, are used to provide the base line for mapping the third gene which is of unknown location and, because it contributes to continuous variation, is not capable - or at least not readily capable - of being followed by a classical methodology.

Consider the situation where a gene A-a, which contributes to continuous variation, is segregating at the same time as two marker genes, G-g and H-h. It should be noted that this time the marker genes are not associated with any inversion, since unhampered recombination is essential in a three-point experiment, being in fact the means by which the location of the new gene is ascertained on the genetic map. There are two situations possible, in the first of which A-a lies between G-g and H-h as shown in the upper part of Fig. 3, and the second in which A-a lies outside the length of chromosome delimited by G-g and H-h as shown in the lower part of Fig. 3. Let the frequency of recombination between A-a and G-g be p_1, that between A-a and H-h be p_2, and that between G-g and H-h be p_3. We will assume that the map distances between the genes are sufficiently small for interference to be complete, i.e. for there to be no double crossing-over within the length of chromosome we are discussing. Each of the recombination values is thus also the frequency of crossing-over.

Let us deal first with the situation where A-a is between G-g and H-h [Fig. 3 (upper)]. Since interference is complete $p_3 = p_1 + p_2$. The triple heterozygote GAH/gah will produce six types of gamete with the frequencies shown in the figure, gametes of types GaH and gAh not being produced because of the absence of double crossing-over. If the triple heterozygote is back-crossed to gah/gah, zygotes of the six corresponding types (GAH/gah, gah/gah, etc.) will be produced with corresponding frequencies. These six genotypes fall into four classes distinguishable by the segregation of the marker genes, namely GH/gh, Gh/gh and gH/gh and gh/gh, but we cannot distinguish between A/a and a/a in the same way since this gene contributes to continuous variation and its segregation is obscured by non-heritable variation and by the effects of any other genes which contribute to the variation of the character and which

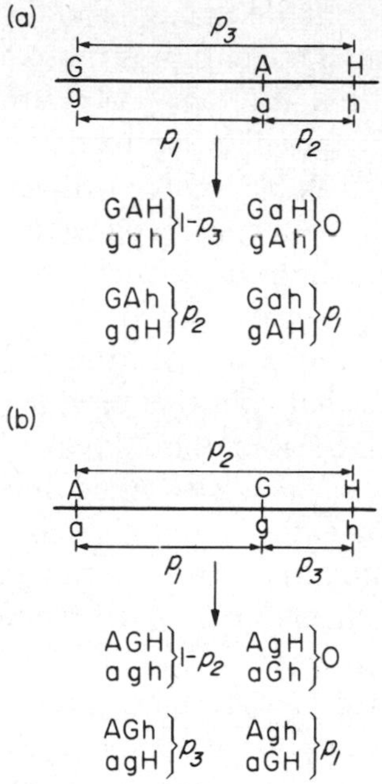

Fig. 3. Locating a gene difference (A-a) affecting continuous variation by reference to two marker genes (G-g and H-h). The gametic output of the triple heterozygote is shown, above, where A-a lies between G-g and H-h, and below, where A-a lies outside the segment delimited by G-g and H-h. p_1 is the frequency of recombination between A-a and G-g; p_2 that between A-a and H-h; and p_3 that between G-g and H-h. Interference is assumed to be complete.

may also be segregating. We can, however, record the average expression of the character in each of the classes GH/gh, Gh/gh, gH/gh and gh/gh. Let Aa add an increment d and aa an increment of $-d$ to the mean expression.

The four classes distinguished by the marker genes are shown in Table 3, together with the frequencies in which they occur associated with A and a respectively, and also their overall frequencies in the progeny, which must of course depend on p_3. (Note that only the genes which these genotypes received from the triple heterozygotes, and by which they are distinguished, are shown in the Table: all individuals received gah from the other parent.) Now all individuals in the marker class GH carry A, and hence will show a mean expression of $m+d$ in respect of the continuously varying character where m is the mean expression of the whole experiment. Similarly the gh class always carries

TABLE 3.

Locating a gene between the markers
(Observed results from Wolstenholme and Thoday, 1963)

Marker class	Frequency			Mean	Observed
	A	a	Joint		
GH	$\frac{1}{2}(1-p_3)$	0	$\frac{1}{2}(1-p_3)$	d	21.16
Gh	$\frac{1}{2}p_2$	$\frac{1}{2}p_1$	$\frac{1}{2}p_3$	$d(p_2-p_1)/p_3$	19.59
gH	$\frac{1}{2}p_1$	$\frac{1}{2}p_2$	$\frac{1}{2}p_3$	$-d(p_2-p_1)/p_3$	18.86
gh	0	$\frac{1}{2}(1-p_3)$	$\frac{1}{2}(1-p_3)$	$-d$	17.86
	$\frac{1}{2}(\overline{GH}-\overline{gh}) = d = 1.650$			$d = 1.650$	
	$\frac{1}{2}(\overline{Gh}-\overline{gh}) = d(p_2-p_1)/p_3 = 0.365$			$p_1 = 0.050$	
	$p_1 + p_2 = p_3 = 0.129$			$p_2 = 0.079$	

a and so has a mean of $m-d$. The Gh class comprises two genotypes: GAh with a frequency of $\frac{1}{2}p_2$ and an expression $m+d$, and Gah with a frequency of $\frac{1}{2}p_1$ and an expression $m-d$. The mean of the Gh individuals will thus be

$$(\tfrac{1}{2}p_2 d-\tfrac{1}{2}p_1 d)/(\tfrac{1}{2}p_2+\tfrac{1}{2}p_1) = d(p_2-p_1)/(p_1+p_2) = d(p_2-p_1)/p_3.$$

The mean of the gH marker class is similarly $-d(p_2-p_1)/p_3$. Now writing $\overline{GH}$ for the mean expression of marker class GH, we can see that

$$\tfrac{1}{2}(\overline{GH}-\overline{gh}) = \tfrac{1}{2}[d-(-d)] = d,$$

$$\text{and } \tfrac{1}{2}(\overline{Gh}-\overline{gH}) = \tfrac{1}{2}d\left[\frac{p_2-p_1}{p_3}-\left(-\frac{p_2-p_1}{p_3}\right)\right] = d\,\frac{p_2-p_1}{p_3}.$$

In addition, p_3 can be found from the frequencies of the four marker classes, and we can thus obtain estimates of d, p_1, p_2 and p_3.

Turning now to the second situation where A-a lies outside the piece of chromosome determined by G-g and H-h, the types of gamete produced by the triple heterozygotes are shown, together with their frequencies in Fig. 3 (lower). Table 4 is obtained from Fig. 3 (lower) in the same way as Table 3 was from Fig. 3 (upper). Again, of course, p_3 can be found from the frequencies of occurrence of the four marker classes, but we see that the estimate of d is yielded not by $\frac{1}{2}(\overline{GH}-\overline{gh})$ but from the recombinant marker classes as $\frac{1}{2}(\overline{Gh}-\overline{gH})$. The difference between the parental marker classes GH and gh, provides an estimate of p_1+p_2, since $\frac{1}{2}(\overline{GH}-\overline{gh}) = d(1-p_1-p_2)/(1-p_3)$ from which p_1+p_2 can be found, as

TABLE 4.
Locating a gene outside the markers

Marker class	Frequency A	Frequency a	Frequency Joint	Mean
GH	$\frac{1}{2}(1-p_2)$	$\frac{1}{2}p_1$	$\frac{1}{2}(1-p_3)$	$d(1-p_1-p_2)/(1-p_3)$
Gh	$\frac{1}{2}p_3$	0	$\frac{1}{2}p_3$	d
gH	0	$\frac{1}{2}p_3$	$\frac{1}{2}p_3$	$-d$
gh	$\frac{1}{2}p_1$	$\frac{1}{2}(1-p_2)$	$\frac{1}{2}(1-p_3)$	$-d(1-p_1-p_2)/(1-p_3)$

$$\tfrac{1}{2}(\overline{Gh}-\overline{gH}) = d$$
$$\tfrac{1}{2}(\overline{GH}-\overline{gh}) = d(1-p_1-p_2)/(1-p_3)$$
$$p_1+p_3 = p_2$$

we have already estimates of d and p_3. Now when A-a is to the left of G-g, $p_2 = p_1+p_3$ giving $p_3 = p_2-p_1$. Then p_1 and p_2 can be estimated as $\frac{1}{2}(p_1+p_2-p_3)$ and $\frac{1}{2}(p_1+p_2+p_3)$ respectively.

We can illustrate this method of locating a gene contributing to continuous variation by reference to data from Wolstenholme and Thoday (1963). These authors report a number of such experiments in *Drosophila melanogaster*, and the results of one of these experiments are set out in the right-hand column of Table 3. The continuously varying character is the number of sternopleural chaetae while the marker genes are clipped wing (cp) and Stubble bristles (Sb), which are located respectively at 45.3 and 58.2 on the standard map of chromosome III. The average number of chaetae for the four marker classes are shown in Table 3, but the authors do not report the frequencies of these classes. A direct estimate of p_3 is thus not available from this experiment, but the marker genes are 12.9 units apart on the standard map, and p_3 may therefore be taken as 0.129.

The first thing is to note that the GH class has the greatest mean number of chaeta and gh the lowest. The gene affecting chaeta number (A-a) must thus lie between the two markers: had it been outside, the G-h and gH classes would have shown the extreme mean chaeta numbers (see Table 4). We then proceed, using the formulae of Table 3 to find $d = \frac{1}{2}(\overline{GH}-\overline{gh}) = \frac{1}{2}(21.16-17.86) = 1.650$ and

$$d(p_2-p_1)/p_3 = \tfrac{1}{2}(\overline{Gh}-\overline{gH}) = \tfrac{1}{2}(19.59-18.86) = 0.365$$

giving
$$p_2-p_1 = \frac{0.365 \times 0.129}{1.650} = 0.0285.$$

With $p_1+p_2=p_3$ we then find $p_1=\frac{1}{2}(0.129-0.0285)=0.050$ and $p_2=\frac{1}{2}(0.129+0.0285)=0.079$.

The experiment thus places the locus of A-a at 0.05 × 100 = 5.0 units to the right of cp and 7.9 units to the left of Sb, that is at locus 50.3 on the standard map of chromosome III.

It has been assumed for the purpose of illustration that the effect on sternopleural chaeta number was acribable to a single gene. In fact Wolstenholme and Thoday obtained evidence that two genes were most probably involved. They used in their analysis a technique, introduced by Thoday (1961), of using progeny tests to ascertain the number of classes genetically different in respect of chaeta number included in each of the marker classes. This method of Thoday's has been used by Davies (1971) to show that genes at a minimum of fifteen loci, scattered over the lengths of all three major chromosomes, are involved in the heritable variation of sternopleural chaeta number in *Drosophila melanogaster*, and that similarly at least fourteen or fifteen loci, not the same as those for sternopleural chaetae, are involved in the variation of abdominal chaeta number in this fly. Further evidence from other experiments of various kinds also indicates that the minimum number of gene loci in the variation each of these two chaeta characters is likely to be nearer 20 than 10.

Summarizing, these experiments with *Drosophila melanogaster* show us that the heritable component of the continuous (or to be more precise, quasi-continuous) variation in both abdominal and sternopleural chaeta number depends on genes which are carried on the chromosomes and which will therefore segregate and recombine in just the same way as the familiar genes of classical genetics. Furthermore, within the technical limitations of the experiments, the whole of this heritable component is accountable in terms of such chromosome-borne genes. Differences in chaeta number may reflect the simultaneous action of genes carried on all of three of the major chromosomes and finer analysis reveals that at least some fourteen or fifteen loci must be involved.

The effects of the different genes supplement one another, their effects sometimes combining in a simple additive fashion, but sometimes interacting in such a way that the combined effect is not simply the sum of the individual actions. At the same time, overlaying the variation due to these genes is variation traceable to environmental agencies or to the vagaries of development, variation which is distinguishable from that due to the genes only by a breeding test. Finally the effects traceable to individual genes, or even to whole chromosomes, may be no greater in magnitude, and indeed may often be smaller than the effects of the non-heritable agencies. In other words, as revealed in these experiments the heritable portion of continuous variation depends on genes transmitted in the Mendelian fashion, but acting in polygenic systems, the member genes of a system having effects similar to one another (and to those of non-heritable agencies), capable of supplementing one another (whether in simply additive fashion or not) and small in relation to the non-heritable variation, or at least in relation to the variation in the system as a whole.

2

The biometrical approach

5. The manifestation of polygenic systems

The evidence that we have examined in the previous chapter showed that continuous variation is partly heritable and partly non-heritable, the two components being separable only by appropriate breeding tests. The non-heritable component springs partly from the impact of differences in external environmental agencies, but it may also reflect vagaries in the internal development of the individuals. The heritable component of the variation, as exemplified in the *Drosophila* experiments, depends on genes at many loci scattered over all the chromosomes, but working together in a polygenic system. Because of their small, similar and supplementary effects on the phenotype of its constituent genes, such a system characteristically gives rise to continuous variation, in which the effects of the individual genes cannot be traced except by using special techniques such as are available in well studied species like *Drosophila.*

Polygenic systems have properties which are basic to our understanding of the genetical structure of populations, their variation and their responses to selection (Mather, 1973). These, however, are not our present concern, which is the genetical analysis of the continuous variation that these systems characteristically produce.

A very simple example of a polygenic system and the variation it produces is illustrated at the top of Fig. 4. Two gene pairs are involved, A-a and B-b, the alleles denoted by capital letters each adding a unit to the expression of the character, and those denoted by small letters each substracting a unit from it. It is assumed that these genes show no dominance, i.e. the expression of a heterozygote, Aa or Bb, is mid-way between those of the corresponding homozygotes, AA and aa or BB and bb. The effects of the genes at the two loci supplement one another in a simply additive fashion and the alleles A and a are equally common as are B and b. The genes at the two loci are assumed to be uncorrelated in their distribution, so that the frequencies shown for the various genotypes are those which would be obtained in an F_2 where the genes are

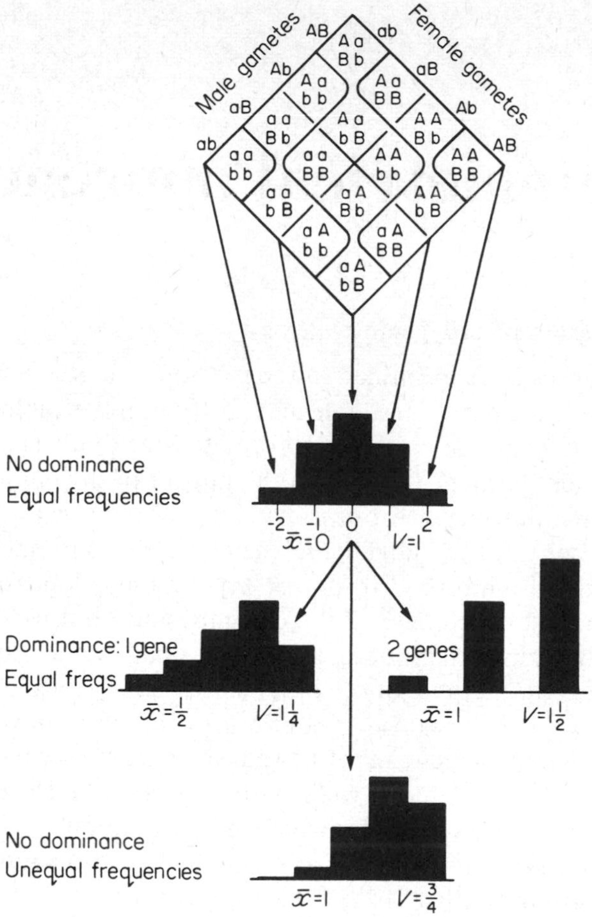

Fig. 4. The polygenic interpretation of continuous variation. The uppermost histogram shows the distribution of phenotypes with two genes of equal and additive effect, and without dominance, neglecting non-heritable variation. The frequencies of alleles A and a, and also of B and b are equal. Each capital letter adds $\frac{1}{2}$ and each small letter $-\frac{1}{2}$ to the phenotypic expression. The two histograms in the centre show the effect of dominance for, on the left, one gene and, on the right, both genes. Dominance is assumed to show itself by the gene denoted by the small letter having no effect when heterozygous with its allele denoted by the capital letter. The histogram at the bottom shows the effect of unequal gene frequencies: the frequencies of A and B are assumed to be $\frac{3}{4}$ and those of a and b to be $\frac{1}{4}$. The mean ($\bar{x}$) and variance (V) are shown below each histogram. In these examples both dominance and unequal gene frequencies produce skewness in the distribution, besides altering the mean and variance.

unlinked. The genic composition of the family is shown at the top of the figure and the distribution of the phenotypes, in the absence of non-heritable differences, is shown immediately below it. Because of the absence of dominance and the simple additivity in their effects of the non-allelic genes, the phenotypic expression of any genotype is proportional to the difference between the numbers of capital letters (denoting alleles enhancing the character) and small letters (denoting alleles diminishing the character). As a consequence certain genotypes give the same phenotype as one another, the most striking example being provided by AaBb, AAbb and aaBB, which all contribute to the central and most common phenotypic class. This similarity of the phenotypes associated with several genotypes combines with the greater frequencies of certain genotypes in the family to produce a frequency distribution in which the central expression is the most common and the extreme expressions most rare, as is characteristic of continuous variation. Since each gene which enhances the character is matched by an equally common allele which diminishes it, the distribution has a mean ($\bar{x}$) of 0, a variance (V) of 1 and is symmetrical.

We can vary the assumptions on which the model is based. Suppose, for example, that we introduce dominance at one of the two loci, say A-a, such that Aa no longer falls on the mid-point between AA and aa but has a phenotype like that of AA. The genotypes occur with the same frequencies as before, but AaBB has a phenotype of 2 like AABB, AaBb joins AABb in having a phenotype of 1, and Aabb joins AAbb and aaBB in having a phenotype of 0, leaving aaBb and aabb with phenotypes of −1 and −2 respectively. The frequency distribution of phenotypes is thus changed to that shown in Fig. 4 (centre left). The mean has been raised from 0 to $\frac{1}{2}$, the variance has increased to $1\frac{1}{4}$, and the distribution is now asymmetrical with the long tail at the lower end. Making both A and B dominant over their respective alleles changes the distribution even more. The mean has risen further to 1 and the variance to $1\frac{1}{2}$ while the asymmetry is now so great that the extreme large phenotype is the most common and certain of the phenotypes have vanished altogether.

Let us now revert to the assumption of no dominance, but alter the gene frequencies so that A and a and B and b are no longer equally common in the population. Let A occur with three times the frequency of a and B with three times that of b, or to put it another way, let the gene frequencies be A $\frac{3}{4}$; a $\frac{1}{4}$ and B $\frac{3}{4}$; b $\frac{1}{4}$. The genotypes will give the same phenotypes as in the original model at the top of Fig. 4, but they will occur with different frequencies. Thus the proportion of AABB individuals will be $\frac{3}{4} \times \frac{3}{4} \times \frac{3}{4} \times \frac{3}{4} = \frac{81}{256}$, that of AaBB and AABb will each be $2 \times \frac{3}{4} \times \frac{1}{4} \times \frac{3}{4} \times \frac{3}{4} = \frac{54}{256}$, and so on. The resulting frequency distribution of phenotypes is shown at the bottom of Fig. 4. In some respects the change in the distribution resembles that brought about by dominance: the mean is again raised to 1 and the distribution is asymmetrical with the long tail towards the lower end. This new distribution differs however from that produced by dominance in that the variance has not been raised but in fact reduced from 1 to 3/4. Thus both the assumptions of dominance and unequal gene frequencies result in change of the biometrical properties of the distribution of phenotypes, and each produces its own characteristic syndrome of changes.

Although broadly resembling the distribution of a continuously varying character, the distributions in Fig. 4 differ from it in one important respect: they are not strictly continuous since the phenotypes fall into a small number of discrete classes. This difference stems from three simplifying assumptions that we have made in the models on which the frequency distributions of Fig. 4 are derived. In the first place we have assumed that the effects of A-a and B-b are alike: had we not made this assumption a larger number of phenotypes would have been possible. Secondly, we have assumed the absence of non-heritable variation: its presence would have blurred the boundaries of the phenotypic classes given by the various genotypes and caused them to overlap, so producing continuous variation. Thirdly, we have been considering a very simple polygenic system comprising only two gene pairs, which when the action of the two gene pairs are alike produces only five phenotypic classes, non-heritable effects apart. The consequences of raising the number of gene pairs in the system are illustrated in Fig. 5. With four loci involved, there are nine phenotypic classes and with eight loci there are seventeen. Thus, given the same overall difference between the extreme phenotypes the step produced by each gene substitution is smaller, and a given change requires more gene substitutions to produce it, the more genes there are in the system. The result is a closer approximation to continuous variation, and although in principle there are small discontinuities still present in the distribution of phenotypes, decreasing amounts of non-heritable variation would serve to blur them and give full continuity.

One further point should be observed about the distribution shown in Fig. 5. All of them are based on the assumptions of no dominance and equal frequencies of the two alleles at each locus. In consequence all the distributions are symmetrical and have means of 0. But the variances of

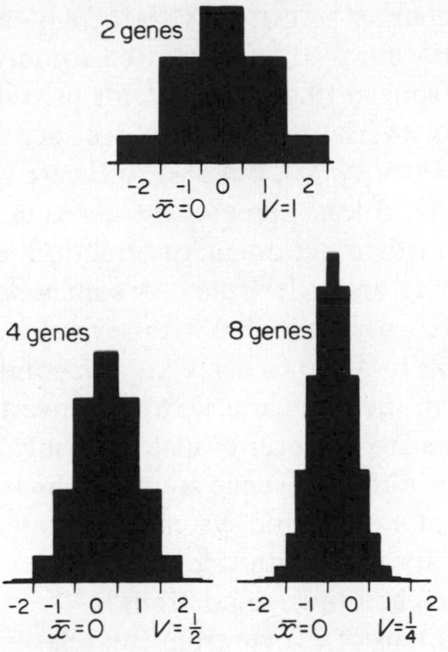

Fig. 5. The effect of change in the number of genes in the polygenic system. The three histograms show the distributions where the systems comprise two, four and eight genes respectively. In all cases the gene frequencies are equal, and the genes in the system have equal and additive effects, without dominance. The range between the highest and lowest expressions of the character is the same in all three cases, the genes in the four gene and eight gene cases thus having individual effects respectively one-half and one-quarter of those of the genes in the two gene case. The number of genotypic, and hence phenotypic, classes rises with the number of genes and the approximation to fully continuous variation becomes closer. The mean of the distribution is unchanged, but the variance falls inversely proportionally as the number of genes rises.

the distributions decrease as the number of gene-pairs increases, that with four gene pairs having half the variance of that with two, and that with eight gene pairs having a quarter of its variance. Again we can see how the genetic properties of the polygenic system are reflected characteristically in the biometrical properties of the frequency distribution of the phenotypes.

6. Genetic analysis and somatic analysis

The use of the special stocks and special breeding methods available in

Drosophila have enabled us not only to recognize that the heritable part of continuous variation is to be attributed to polygenic systems (and indeed it was recognized that such systems provide a basis for understanding continuous variations, long before such experiments were undertaken with *Drosophila*), but also to locate within the chromosomes, and hence to count, at least some of the genes in the system and to investigate up to a point their action and interaction in producing their effects on the chaeta characters under study. A somewhat similar although less detailed analysis has been possible with several characters in wheat, again using special stocks built up to carry known combinations of chromosomes derived from the two varieties under investigation (Law, 1967). Such special stocks are however available in only a limited number of species. How then in their absence are we to proceed to learn something of the properties of a polygenic system, its properties of dominance and the interaction of its genes with one another and with non-heritable agencies, as well as their linkage relations?

The difficulty stems of course from the relatively small effects of individual genes on the character, from the similarity of these effects and from the obscuring effect of the non-heritable portion of the variation. The classical technique of genetics would be to isolate as many of the genes as possible and to study their properties individually, and this is what has been done at any rate up to a point with *Drosophila* using their specific locations on the chromosomes as the basis for recognizing them as separate genes. In principle, while not being able to assign the individual genes to specific locations on specific chromosomes we could proceed some distance in this way with any example of continuous variation. We could seek to control the environment in which the organism is raised so as to reduce the non-heritable variation and its blurring effect on genic segregation, although in so far as the non-heritable differences arose from chance effects of development rather than from the impact of outside agencies, this variation could not be wholly eliminated. We could seek to produce inbred lines from the population or the descendants of the cross under investigation so as at least partially to break down the polygenic system into smaller elements, depending on fewer gene differences, and to provide ourselves with the means of making repeatable observations and progeny tests to whatever extent was necessary to establish a genetic difference however small it might be. Many such inbred lines would be needed, and in the absence of special stocks many generations of inbreeding to give us the material we needed. And in the end when we came to put the parts of the polygenic system together again in order to see how they interacted we should be faced once more with much of the genetic complexity that we had been seeking to circumvent. Such an approach is clearly not generally a practical proposition.

A different approach to fractionating the polygenic system has been advocated from time to time, that of analysing the character under study into component sub-characters in the expectation that these sub-characters would prove to be under simpler, and hence more readily analysable, genetic control than the full character itself. Thus the yield of grain of a wheat plant can be regarded as the product of the average weight per grain, the average number of grains per ear, and the number of ears borne by the plant. If different genes mediated these separate sub-characters we should then have at least made a start on simplifying the problem of genetically analysing the continuous variation in yield, especially if we could at the same time reduce the non-heritable component of the variation.

On the face of it, there are some grounds for believing that this approach through what has been called somatic analysis, might have value as an aid to the genetic analysis. It has been reported by Spickett (1963) that he was able to identify genes in *Drosophila* by their location in the chromosomes, all of which affected the number of sternopleural chaetae and did so in different ways, one by a local effect in a particular section of the clump of chaetae while another had a more generally distributed effect. At a somewhat coarser level, genes are known which affect the number of sternopleurals while not affecting the number of abdominal chaetae and vice versa. But other genes are also known which affect both sets of chaetae simultaneously. These genes can be recognized by effects other than on the sternopleurals and abdominals and they are genes producing discontinuity in the distribution of phenotypes and so capable of being followed by the Mendelian technique. But if their effects were confined to the chaetae under consideration and were sufficiently small not to produce individually detectable characteristics, and if they were segregating simultaneously in a family or population we should find that seeking to analyse the genetic control of the one group of chaetae separately from that of the other did not in fact simplify the problem; for while this somatic analysis would serve to separate some of the genes it would not separate others which affected both sub-characters and which therefore appeared in both analyses. Variation in the two sub-characters would be correlated because some genes affected both, but only partially so because other genes affected only one.

This situation is a commonplace in Mendelian genetics. Taking but one example, that of flower colour in plants, genes are known which simultaneously affect both the anthocyanin and anthoxanthin pigments, others which affect only the one class of pigment and still further genes which affect only the other. Even these latter genes can result in correlated effects in the two classes of pigment, for the two types of pigment can share a common precursor which if in limited supply will be available in greater quantity for the production of one type if, because of gene action, the other type is being produced in lesser quantity and so is making smaller demands on the pool of precursor. Thus a negative correlation can arise between the amounts of the two kinds of pigment.

The evidence from attempts at the somatic analysis of continuously varying characters agrees with this expectation. If we subdivide yield of grain in a cereal into average weight of grain, average number of grains per ear, and number of ears in the plant, we find that there are correlations, most commonly negative ones, between the sub-characters. Similarly the yield of sugar by sugar beet is the product of the sugar percentage in the root and the weight of root; but the two are negatively correlated and while it is relatively easy to raise the yield of root by selection, the sugar percentage will then tend to fall and vice versa. In seeking to breed for yield of sugar little advantage is gained by treating the two subcharacters separately, for the value of this somatic analysis is largely vitiated by the negative correlation between them.

At the fundamental level of the gene and its immediate biochemical product, there can be a simple one-to-one correspondence between change in the gene and change in the product, as indeed we see in the variation of such proteins as haemoglobins and enzymes. But when we pass to characters of the kind we have been discussing, biochemically and developmentally remote from the primary action of the genes, the complexity of development ensures that just as the character will be affected by many genes, one gene may - and indeed commonly will - be found to affect a number of characters, if we search out all its consequences for the overall phenotype of the organism. So, save at a very basic level, somatic analysis and genetic analysis will not march together in a simple fashion, and the only way to relate changes of phenotype to changes of genotype is to isolate the genes and ascertain their effects. Somatic analysis is of use only after it has been validated by prior genetical analysis: it is not a generally reliable precursor to genetic analysis itself. It is of use for genetical purposes only where experiment and observation have shown its application to be justifiable and helpful: where we are dealing with continuous variation, due to genes which in general we cannot expect to be readily recognizable in segregation, we cannot expect to overcome the intrinsic difficulty of the situation by attempting a prior somatic analysis.

7. Biometrical genetics

If we accept that commonly we cannot distinguish any of the individual genes whose segregation contributes to continuous variation (and that even with the special stocks available in *Drosophila* we cannot distinguish all of them) we must be content to deal with the relevant polygenic system as a whole. And since we cannot distinguish the segregant classes one from another, we cannot use a form of analysis based on class frequencies as in the classic Mendelian method. We can, however, recognize the biometrical properties of the frequency distributions of the phenotypes which are our raw material, and we can estimate the biometrical quantities, the means, variances, and so on, which characterize these distributions. As we have seen, these parameters can reflect, and reflect in characteristic ways, the properties of the polygenic system from which the heritable component of the variation stems. We can thus seek to gain information about the properties of the genes underlying continuous variation by analysis of the biometrical quantities which characterize the frequency distributions of the phenotypes in related families and populations. We must expect that the information so obtained will not be just like that yielded by classical genetical analysis. In particular, since we shall not be following individual genes we cannot learn about their individual properties: rather, since we are considering the system as a whole, we shall obtain information about the overall joint or average properties of its member genes. At the same time because we are considering all the variation that the character shows we shall be bringing the effects of all the relevant genes into the reckoning, and this we can never achieve by the Mendelian technique of identifying and following individual genes, since there must inevitably be some genes of relatively small effect which escape identification.

The phenotypes of the individuals in any family or other appropriate group yield two biometrical quantities which are of use to us, the mean of the distribution (a first degree statistic since it is linear in x, the metric measuring the expression of the character) and the variance (a second degree statistic depending on x^2). In addition, any pair of related families or groups may yield a covariance, which is also a second degree statistic. Higher order statistics may also be obtained, notably that of the third-order which measures skewness (depending on x^3) and the fourth order which measures kurtosis (depending on x^4). These have, however, seldom been put to use in genetical analysis and we shall consider them no furthe. We shall thus be concentrating on the genetical information that can be derived from comparisons among the means, variances and covariances of related families or groups of individuals. These we shall seek to interpret in terms of appropriate parameters representing the consequences of the various genetical phenomena in which we may be interested. Having defined these parameters, expectations are formulated in terms of them for the means, variances and covariances of the families or groups that our experiments yield. The means, etc. observed are then related to these expectations in such a way as to yield estimates of the parameters and tests of their significance.

In any experiment we may run into a complexity of genetical phenomena, especially as we must expect to be dealing with a number of genes whose relations one with another may not be the same for all of them: indeed we have already seen this to be the case with the system mediating variation of the number of sternopleural chaetae in *Drosophila*, where the genes of the X chromosome interacted with those of chromosomes II and III, although these latter show no evidence of anything but an additive relation to one another. Such a complexity of phenomena leads to a corresponding multiplicity of parameters which it would be necessary to take into account in formulating expectations for the statistics observed, with the consequence that except in large and complex experiments there could be more parameters than there were statistics from which to estimate them. Some simplification must therefore be made in the approach: only those parameters which are regarded as of chief importance, and with which the data can cope, should be introduced into the analysis initially, and others added only as necessity requires.

The simplest genetical formulation to be used in the initial analysis is generally taken as that which includes parameters representing the additive effects of the genes (that is the differences between corresponding homozygotes, AA and aa, BB and bb, etc.) and their dominance properties. Given that the experimental material is sufficient, the experiment adequately designed and the statistical analysis suitably carried out, we can then estimate these parameters and also test the goodness of fit of this initial simple formulation to the observations. If the fit proves to be adequate, we have no grounds for postulating a more complex genetical situation. But if, on the other hand, the fit proves to be inadequate, consideration can be given to a more complex formulation incorporating further parameters, representing interaction between non-allelic genes, or linkage or whatever else seems appropriate. If this in turn proves to be inadequate to fit the observations, and the data are themselves sufficiently extensive, a still more complex set of parameters representing a still more complex genetical situation can be tried.

This approach will be developed and illustrated in the following chapters. We shall start by considering data from controlled breeding experiments based on crosses among true-breeding lines and later turn to the more difficult analysis of data from randomly breeding populations, just as classical genetics began with experimental crosses and later proceeded to the genetical analysis of populations.

3

Additive and dominance effects

8. Components of means

With disomic inheritance, two alleles A-a can give rise to three genotypes AA, Aa and aa. Two parameters are required to describe the differences in phenotypic expression of these three genotypes in respect of any character which they affect. As the origin, we take the mid-point between the two homozygotes since this does not depend on the differences between the three genotypes, but on the rest of the genotype and the effects of the environment, and thus reflects the general circumstances of the observations. The two parameters measuring the differences between the genotypes may then be defined as d, measuring the departure of each homozygote from the mid-point, and h, measuring the departure of the heterozygote from it. Taking A as the allele which increases the expression of the character, AA will exceed the mid-point (m) by d, and so will have an expression $m + d$, while aa will equally fall short of the mid-point having an expression $m-d$, and Aa will deviate from m by h so having an expression $m + h$ (Fig. 6). If h is 0 the hetero-

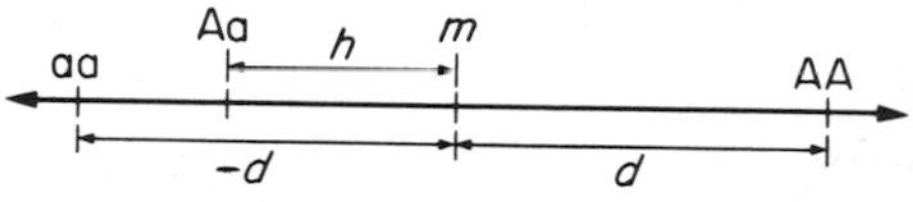

Fig. 6. The d and h increments of the gene difference A-a. Deviations are measured from the mid-parent, m, midway between the two homozygotes AA and aa. Aa may lie on either side of m and the sign of h will vary accordingly.

zygote's expression of the character will be midway between the expression of the two homozygotes and dominance is absent. If h is positive, the heterozygote will be nearer to AA than to aa in its expression and A will be partially, or if $h = d$ completely, dominant. Similarly if h is negative, a will be the dominant allele. If $h > d$ Aa will fall outside the range delimited by AA and aa, and the gene may then be said to display over-dominance. It should be noted that here the capital letter A does not imply dominance of the allele so designated: A is the allele which increases the expression of the character whether it be dominant or not.

This characterization of the differences among the genotypes can be applied to any genes, whether their effects be large or small, leading to continuous variation or not, provided the expressions of the character in question can be expressed in quantitative terms. Thus the sex-linked mutant Bar-eye (B) reduces the number of facets in the eyes of *Drosophila melanogaster*, wild-type females (+/+) having an average number of 779.4 facets, heterozygotes (B/+) having an average of 358.4 facets and the homozygous mutant (B/B) having an average of 68.1 at 25°C (Sturtevant, 1925, quoted by Goldschmidt, 1938). Then m is $\frac{1}{2}(779.4 + 68.1) = 423.75$, $d = 779.4 - 423.75 = \frac{1}{2}(779.4 - 68.1) = 355.65$ and $h = 358.4 - 423.75 = -65.35$. Since h is negative the B mutant is partially dominant to wild-type and we may if we wish measure its degree of dominance by $h/d = -65.35/355.65 = -0.184$. We should note that the effect of the Bar-eye mutant is large, and leads to discontinuous variation, the phenotypes of B/B, B/+ and +/+ showing no overlap. No one would go to the trouble of counting the facets in classifying the three genotypes when Bar-eye is being used, and because its effect is sufficiently large for it to be recognized and followed individually in breeding experiments there would be no difficulty in disentangling it from other gene differences whose effects were sufficiently small to contribute only to the continuous variation in facet number that we can observe within the phenotypes associated with each of these genotypes for Bar-eye.

Confining ourselves now to continuous variation, we cannot distinguish individually the genes contributing to it. If we consider two homozygous lines the departure of each of them from their mid-point (or mid-parent as it is often called) will reflect the simultaneous action of all the genes affecting the character by which the lines differ. Assuming that the effects of these genes are simply additive, the departure from the mid-point will in fact be the sum of the d's, one from each of the genes, taking sign into account. Where, for example, the lines differ at two loci, A-a and B-b, if one of them is AABB and the other aabb, the first will depart by $d_a + d_b$ and the second by $-(d_a + d_b)$. But if the lines are AAbb and aaBB, their departures will be $d_a - d_b$ and $-d_a + d_b$ respectively. Generalizing, where the homozygous lines differ at k

loci, we may define $[d]$ as the departure from the mid-parent of the line with the greater expression of the character, where $[d] = S(d_+) - S(d_-)$, $S(d_+)$ standing for the sum of the d's of all the genes in this line tending to increase the phenotype, $S(d_-)$ for the sum of the d's of those tending to decrease it and $S(d_+) > S(d_-)$ since $[d]$ must be positive. In the same way, when we cross the two homozygous lines, the phenotype of the heterozygote will depart from the mid-parent by $[h] = S(h)$. Since by definition any h may be positive or negative, $[h]$ itself may be positive or negative, and of course where some of the genes at some of the loci have positive h's and others negative h's they will tend to balance out each other's effects. $[h]$ may thus be small or even 0, even where each of the genes individually shows pronounced dominance, simply because being dominant in opposite directions they are cancelling out each other's effects.

We can now see at once that although h/d provides a measure of dominance for a single gene difference, $[h]/[d]$ does not provide a corresponding measure of dominance when we are considering more than one gene. $[h]/[d]$ may be very small simply because some of the h's are positive and others negative, so leading to a small value for $[h]$ even although none of the individual h's is small; and equally $[h]/[d]$ may be large just because the genes are so distributed between the parent lines that they are tending to balance out one another's effects and $[d] = S(d_+) - S(d_-)$ is small even although every d is itself not small. Thus $[h]/[d]$, although depending on dominance in that it cannot depart from 0 unless one or more of the genes show dominance, is not itself a direct measure of that dominance. For this reason it is often referred to as the potence ratio. It is particularly worth emphasizing that where the F_1 between two lines differing at more than one locus gives a phenotype falling outside the range delimited by the parents and so displays heterosis, i.e. $[h]>[d]$; there is no reason to postulate over-dominance of any of the genes involved since the excess of $[h]$ over $[d]$ can come about merely by the d's of the various genes balancing one another to a greater extent than do their h's. Thus to take a simple example, when $h_a = d_a$ and $h_b = d_b$, the F_1 between AAbb and aaBB will have a phenotype of $h_a + h_b$, the parents having phenotypes of $d_a - d_b$ and $-d_a + d_b$. Then $[h]/[d] = (h_a + h_b)/(d_a - d_b) = (d_a + d_b)/(d_a - d_b)$ and heterosis is displayed even although neither gene shows over-dominance.

Where an F_2 is raised from the F_1, it will include $\frac{1}{4}$AA, $\frac{1}{2}$Aa and $\frac{1}{4}$aa in respect of the gene A-a. This gene will therefore contribute $\frac{1}{4}d_a + \frac{1}{2}h_a - \frac{1}{4}d_a = \frac{1}{2}h_a$ to the departure of the average expression of the character in

F_2 from the mid-parent. Assuming the effects to be additive of the k genes by which the parent lines differed, the departure of the F_2 mean thus becomes $\frac{1}{2}[h]$, and it may be observed that this is equally the case even where two or more of the genes are linked. The mean phenotype of the F_2 will then be $\bar{F}_2 = m + \frac{1}{2}[h]$. In the same way, where B_1 is the back-cross to the larger parent P_1, it will include $\frac{1}{2}$AA and $\frac{1}{2}$Aa and A-a contributes $\frac{1}{2}d_a + \frac{1}{2}h_a$ to the departure of the mean of B_1 from the mid-parent. Then taking all k genes into account $\bar{B}_1 = m + \frac{1}{2}[d] + \frac{1}{2}[h]$. Similarly the back-cross to P_2, the smaller parent gives $\bar{B}_2 = m - \frac{1}{2}[d] + \frac{1}{2}[h]$.

Continuing from the F_2, where a true F_3 generation is raised by selfing the F_2 individuals, in respect of A-a it will comprise $\frac{3}{8}$ AA, $\frac{1}{4}$ Aa and $\frac{3}{8}$ aa when taken as a whole. This gene will then contribute $\frac{3}{8}\,d_a + \frac{1}{4}\,h_a - \frac{3}{8}\,d_a = \frac{1}{4}\,h_a$ to the departure of the F_3 mean from the mid-parent, and taking all k genes into account the mean phenotype will be $\bar{F}_3 = m + \frac{1}{4}[h]$. If however the third generation is raised by mating together pairs of individuals taken at random from the F_2 (a procedure which is sometimes incorrectly described, especially by animal geneticists, as giving an F_3 generation) the distribution of A-a over this generation taken as a whole will be $\frac{1}{4}$ AA, $\frac{1}{2}$Aa, $\frac{1}{4}$aa as in the F_2, and the mean phenotype will be $\bar{S}_3 = m + \frac{1}{2}[h]$ where S_3 indicates the third generation raised by sibmating among the F_2. This formulation of mean phenotypes in terms of m, $[d]$ and $[h]$ can be extended to the F_4, where $\bar{F}_4 = m + \frac{1}{8}[h]$, and indeed to any of the types of family raised by the almost endless combinations of mating systems possible among the descendants of the initial cross. A number of these results are collected together in Table 5.

9. Testing the model

We can thus arrive at a formulation of the mean phenotypes in terms of the mid-parent, m, which depends on the general conditions of the observations, the additive component $[d]$ and the dominance component $[h]$. If this formulation is adequate, Table 5 shows that a number of relations must hold good. Thus confining ourselves to the parents, P_1 and P_2, the F_1, the F_2 and the two back-crosses, B_1 and B_2, we can see that

$$\bar{B}_1 = \tfrac{1}{2}(\bar{F}_1 + \bar{P}_1)$$

$$\bar{B}_2 = \tfrac{1}{2}(\bar{F}_1 + \bar{P}_2)$$

$$\text{and} \quad \bar{F}_2 = \tfrac{1}{4}(2\,\bar{F}_1 + \bar{P}_1 + \bar{P}_2).$$

TABLE 5.

Components of means

Generation	Mean Phenotype		
	m	$[d]$	$[h]$
P_1	1	1	0
P_2	1	−1	0
F_1	1	0	1
F_2	1	0	$\frac{1}{2}$
B_1	1	$\frac{1}{2}$	$\frac{1}{2}$
B_2	1	$-\frac{1}{2}$	$\frac{1}{2}$
F_3	1	0	$\frac{1}{4}$
F_4	1	0	$\frac{1}{8}$
S_3	1	0	$\frac{1}{2}$
S_4	1	0	$\frac{3}{8}$
$F_2 \times P_1$	1	$\frac{1}{2}$	$\frac{1}{2}$
$F_2 \times P_2$	1	$-\frac{1}{2}$	$\frac{1}{2}$
$F_2 \times F_1$	1	0	$\frac{1}{2}$
B_1 selfed	1	$\frac{1}{2}$	$\frac{1}{4}$
B_2 selfed	1	$-\frac{1}{2}$	$\frac{1}{4}$

These expected relationships can be used to test the adequacy of the model. The families must have been raised in comparable environments, so that differences between their means which spring from differences of the environments in which they have been raised do not introduce distorting biases into the estimates of the mean phenotypes, $\bar{P}_1$, $\bar{P}_2$, $\bar{F}_1$, $\bar{F}_2$, $\bar{B}_1$ and $\bar{B}_2$. Also these means will be subject to sampling variation which can be estimated by normal statistical procedures from the variances among the individuals within the families themselves. Thus if V_{P1} is the variance of the individuals within the P_1 family, and $V_{\bar{P}1}$ is the variance of $\bar{P}_1$, the mean of P_1, $V_{\bar{P}1} = V_{P1}/n$ where n is the number of individuals observed in P_1 and used in calculating $\bar{P}_1$.

Now we can rewrite the first of the relations as $A = 2\bar{B}_1 - \bar{P}_1 - \bar{F}_1 = 0$ whereupon we can find $V_A = 4V_{\bar{B}1} + V_{\bar{P}1} + V_{\bar{F}1}$ and the standard error of A can be obtained as $\sqrt{V_A}$. The expected value of A is 0 and we can thus test whether this relation holds good by finding $A/\sqrt{V_A}$ and looking up its probability in a table of normal deviates in the customary way. It should be noted that if the numbers of individuals observed within each of the three families, P_1, F_1 and B_1, are small (say less than 10) $A/\sqrt{V_A}$ must be treated as t and its probability found from the table of t using as the number of degrees of freedom the sum of the numbers of df from the three families. The other two relations can similarly be tested by setting $B = 2\bar{B}_2 - \bar{P}_2 - \bar{F}_1$ with correspondingly $V_B = 4V_{\overline{B2}} + V_{\overline{P2}} + V_{\overline{F1}}$ and $C = 4\bar{F}_2 - 2\bar{F}_1 - \bar{P}_1 - \bar{P}_2$ with $V_C = 16V_{\overline{F2}} + 4V_{\overline{F1}} + V_{\overline{P1}} + V_{\overline{P2}}$. These tests of the expected relationship have been termed 'scaling tests' by Mather (1949) and further scaling tests can be devised where observations on additional types of family are available. Thus, for example, where observations have also been made on the F_3 generation we can test the agreement of the relation $D = 8\bar{F}_3 - 3\bar{P}_1 - 3\bar{P}_2 - 2\bar{F}_1$ with its expected value 0 (see Table 5), using $V_D = 64V_{\overline{F3}} + 9V_{\overline{P1}} + 9V_{\overline{P2}} + 4V_{\overline{F1}}$.

Sets of such scaling tests can be devised to cover any combination of types of family that may be available. Instead, however, of testing the various expected relationships one at a time, a procedure proposed by Cavalli (1952) and known as the joint scaling test may be used. This effectively combines the whole set of scaling tests into one and thus offers a more general, more convenient, more adaptable and more informative approach. It consists of estimating the model's parameters, m, $[d]$ and $[h]$ from the means of the all types of families available, followed by a comparison of these means as observed with their expected values derived from the estimates of the three parameters. This makes it clear at once that at least three types of family are necessary if the parameters of the model are to be estimated, but with only three types of family available no test can be made of the goodness of fit of the model since in such a case a perfect fit must be obtained between the observed means and their expectations from the estimates of the three parameters. So to provide such a test at least four types of family must be raised.

The procedure of the joint scaling test may be illustrated by reference to data supplied by Dr D. S. Virk of a cross between two pure-breeding varieties, 22 and 73, of the Birmingham collection of *Nicotiana rustica* varieties. In Table 6 are presented the means and variances of the means for plant height of the parental, F_1, F_2 and first back-cross families (B_1 and B_2) derived from this cross, when grown in the summer of 1975. Family size was deliberately varied with the kind of family. It was set at as low as 20 for the genetically uniform parents and in excess of 100 for the F_2 and back-crosses, to compensate for the greater variation expected in these segregating families. All plants were individually randomized at the time of sowing so that the variation within families reflects all the non-heritable sources of variation to which the experiment is exposed. With this design the estimate of variance of a family mean ($V_{\bar{x}}$) valid for use in the joint scaling test is obtained in the usual way by dividing the

TABLE 6.

Joint scaling test on a cross between true-breeding varieties 22 and 73 of *Nicotiana rustica* for the character final height of the plant in cm

Generation	No. of plants	$V_{\bar{x}}$	Weight $(=1/V_{\bar{x}})$	Model m	$[d]$	$[h]$	Mean Observed	Expected	Difference O-E
P_1 (var 22)	20	1.0034	0.967680	1	1	0	= 116.3000	115.5217	0.7783
P_2 (var 73)	20	1.4525	0.668847	1	−1	0	= 98.4500	99.1223	−0.6723
F_1	60	0.9699	1.031034	1	0	1	= 117.6750	117.3807	0.2943
F_2	160	0.4916	2.034174	1	0	$\frac{1}{2}$	= 111·7781	112.3514	−0.5733
B_1	120	0.4888	2.045827	1	$\frac{1}{2}$	$\frac{1}{2}$	= 116.0000	116.4512	−0.4512
B_2	120	0.6135	1.629992	1	$-\frac{1}{2}$	$\frac{1}{2}$	= 109.1610	108.2515	0.9095

$\chi^2_{[3]} = 3.411$

variance within the family (V_x) by the number of individuals in that family (Table 6). Reference to this table shows that the greater family size of the segregating generations has more than compensated for their greater expected variability in that the variances of their family means are smaller than those of their non-segregating families.

Six equations are available for estimating m, $[d]$ and $[h]$ and these are obtained by equating the observed family means to their expectations, in terms of these three parameters, which are taken from Table 5. The coefficients of m, $[d]$ and $[h]$ in the six equations are listed in the central columns of Table 6. There are three more equations than unknowns and the estimation of the three unknowns (m, $[d]$ and $[h]$) must therefore be by a least squares technique. The six generation means to which we are fitting the m, $[d]$ and $[h]$ model are not known with equal precision; for example, the variance of the mean ($V_{\overline{P2}}$) of $\bar{P}_2$ is almost three times that of the $\bar{B}_1$. The best estimates will be obtained, therefore, if the generation means and their expectations are weighted, the appropriate weights being the reciprocals of the variances of the means. For the first entry in the table, P_1, the weight is given by $1/1.0334 = 0.9677$ and so on for the other families (Table 6).

The six equations and their weights may be combined to give three equations whose solution will lead to weighted least squares estimates of m, $[d]$ and $[h]$, as follows. In order to obtain the first of these three equations each of the six equations is multiplied through by the coefficient of m which it contains, and by its weight, and the six are then summed. We thus have

m		$[d]$	$[h]$		
0.967 680 0	+	0.967 680 0		=	112.541 184 0
0.668 846 8	−	0.668 846 8		=	65.847 967 4
1.031 034 0			+ 1.031 034 0	=	121.326 925 9
2.034 174 0			+ 1.017 087 0	=	227.376 104 8
2.045 826 5	+	1.022 913 3	+ 1.022 913 2	=	237.315 874 0
1.629 991 8	−	0.814 995 9	+ 0.814 995 9	=	177.931 534 9
8.377 553 1	+	0.506 750 6	+ 3.886 030 1	=	942.339 591 0

The second and third equations are found in the same way using the coefficient of $[d]$ for the second and of $[h]$ for the third along with the weights as multipliers. We then have three simultaneous equations, known as normal equations, that may be solved in a variety of ways to yield estimates of m, $[d]$ and $[h]$.

A general approach to the solution is by way of matrix inversion. The three equations are rewritten in the form

$$\underset{\mathbf{J}}{\begin{bmatrix} 8.377\,553\,1 & 0.506\,750\,6 & 3.886\,030\,1 \\ 0.506\,750\,6 & 2.555\,481\,4 & 0.103\,958\,7 \\ 3.886\,030\,1 & 0.103\,958\,7 & 2.458\,532\,1 \end{bmatrix}} \underset{\hat{\mathbf{M}}}{\begin{bmatrix} \hat{m} \\ \hat{d} \\ \hat{h} \end{bmatrix}} = \underset{\mathbf{S}}{\begin{bmatrix} 942.339\,591\,0 \\ 76.385\,386\,1 \\ 442.638\,682\,7 \end{bmatrix}}$$

where $\mathbf{J}$ is the information matrix, $\hat{\mathbf{M}}$ is the estimate of the parameters and $\mathbf{S}$ is the matrix of the scores.

The solution then takes the general form $\hat{\mathbf{M}} = \mathbf{J}^{-1}\mathbf{S}$ where $\mathbf{J}^{-1}$ is the inverse of the information matrix and is itself a variance-covariance matrix.

The inversion may be achieved by any one of a number of standard procedures (Fisher, 1946; Searle, 1966). For our example, inversion leads to the following solution.

$$\underset{\mathbf{M}}{\begin{bmatrix} \hat{m} \\ \hat{d} \\ \hat{h} \end{bmatrix}} = \underset{\mathbf{J}^{-1}}{\begin{bmatrix} 0.456\,785\,3 & -0.061\,314\,0 & -0.719\,416\,0 \\ -0.061\,314\,0 & 0.400\,220\,1 & 0.079\,991\,4 \\ -0.719\,416\,0 & 0.079\,991\,4 & 1.540\,495\,1 \end{bmatrix}} \underset{\mathbf{S}}{\begin{bmatrix} 942.339\,591\,0 \\ 76.385\,386\,1 \\ 442.638\,682\,7 \end{bmatrix}}$$

The estimate of m is then

$$\hat{m} = (0.456\,785\,3 \times 942.339\,591\,0) - (0.061\,314\,0 \times 76.385\,386\,1) - (0.719\,416\,0 \times 442.638\,682\,7)$$

$$= 107.322\,036\,2$$

which equals 107.3220 to the accuracy required, and the S.E. of $\hat{m}$ is $\sqrt{0.456\,785\,3} = \pm 0.675\,859 = \pm 0.6759$ to the accuracy required. In a similar way

$$[\hat{d}] = 8.1997 \pm 0.6326$$

$$\text{and} \quad [\hat{h}] = 10.0587 \pm 1.2412.$$

All are highly significantly different from zero when looked up in a table of normal deviates.

The adequacy of the additive-dominance model may now be tested by predicting the six family means from the estimates of m, $[d]$ and $[h]$. For example,

$$\bar{B}_2 = m - \tfrac{1}{2}[d] + \tfrac{1}{2}[h]$$

on the basis of this model and for the estimates obtained it has as the expected value

$$107.3220 - \tfrac{1}{2}(8.1997) + \tfrac{1}{2}(10.0597) = 108.2515.$$

This expectation along with those for the other five families is listed in Table 6. The agreement with the observed values appears to be very close and in no case is the deviation more than 0.83% of the observed value. The goodness of fit of this model can be tested statistically by squaring the deviation of the observed from the expected value for each type of family and multiplying by the corresponding weight. The sum of the products over all six types of families is a χ^2. Since the data comprise six observed means, and three parameters have been estimated, this χ^2 has $6 - 3 = 3$ degrees of freedom.

The contribution made to the χ^2 by $\bar{P}_1$, for example, is $(116.3000 - 115.5217)^2 \times 0.967\,68 = 0.5862$. Summing the six such contributions, one from each of the six types of family, gives $\chi^2_{[3]} = 3.4110$ which has a probability of between 0.40 and 0.30. The model must therefore be regarded as adequate: there is no evidence of anything beyond additive and dominance effects.

The individual scaling tests, A, B and C, referred to on page 37 can, of course, also be used to test the model. Thus with the present data

$$A = 2\bar{B}_1 - \bar{P}_1 - \bar{F}_1 = (2 \times 116.000) - 116.300 - 117.6750 = -1.975$$

$$\text{and} \quad V_A = 4V_{\bar{B}1} + V_{\bar{P}1} + V_{\bar{F}1} = (4 \times 0.4888) + 1.0334 + 0.9699 = 3.959$$

$$\text{leading to} \quad s_A = \sqrt{V_A} = 1.990.$$

Thus $A = -1.98 \pm 1.99$ which, when entered in a table of normal deviates does not differ significantly from the value 0 expected. These three tests, as applied to the present data, are summarized in Table 7. Not surprisingly they agree with the joint scaling test in showing the model to be adequate.

TABLE 7.

Individual scaling tests on the data from a cross in *Nicotiana* used in Table 6

Test		
$A = 2\bar{B}_1 - \bar{P}_1 - \bar{F}_1$	=	-1.98 ± 1.99
$B = 2\bar{B}_2 - \bar{P}_2 - \bar{F}_1$	=	2.20 ± 2.21
$C = 4\bar{F}_2 - 2\bar{F}_1 - \bar{P}_1 - \bar{P}_2$	=	-2.99 ± 3.77

The joint scaling test, however, does more than test the adequacy of the additive-dominance model: it provides the best possible estimates of all the parameters required to account for differences among family means when the model is adequate and, as we shall see in Chapter 5, it can be readily extended to more complex situations. In the present case, these best estimates show that the additive and dominance components are of the same order of magnitude and since $[h]$ is significantly positive, alleles which increase final height must be dominant more often than alleles which decrease it.

In this example the simple model is adequate but this is frequently not the case, the inadequacy being revealed both by the joint scaling test leading to a significant χ^2 and by one or more of the individual scaling tests showing a significant departure from 0. Two examples of this analysed in the way just described are summarized in Table 8.

The first is the weight per loculus of fruit in a cross between the two tomato varieties, Danmark and Red Currant grown in 1938 (Powers, 1951). The second example, again provided by Dr D. S. Virk, is plant height at the sixth week after planting in the experimental field in a cross between varieties 72 and 22 of *Nicotiana rustica.* Variety 22 was a parent of the cross we have just analysed in detail and 72 has the same origin as variety 73 of the earlier cross. Both crosses were grown simultaneously, using the same experimental design and family sizes, in 1975.

For the tomato cross all three individual scaling tests are significant as is also the joint scaling test. For the *N. rustica* cross the C scaling test

TABLE 8.

Examples of crosses where the additive-dominance model is inadequate.
1. Tomato: Danmark X Red Currant, for weight per loculus of fruit, in 1938 (Powers, 1951)
2. *Nicotiana rustica*: varieties 72 X 22, for plant height at sixth week in field, in 1975.

Generation	Mean and its S.E. Cross 1	Cross 2
P_1	10.36 ± 0.581	80.40 ± 1.936
P_2	0.45 ± 0.017	65.47 ± 1.726
F_1	2.33 ± 0.130	85.99 ± 1.231
F_2	2.12 ± 0.105	84.03 ± 0.856
B_1	4.82 ± 0.253	84.18 ± 1.160
B_2	0.97 ± 0.045	73.88 ± 1.015
Scaling tests		
A	−3.05 ± 0.791	1.97 ± 3.263
B	−0.85 ± 0.159	−3.70 ± 2.936
C	−6.99 ± 0.763	18.27 ± 4.950
Joint	$\chi^2_{[3]} = 96.59$	$\chi^2_{[3]} = 24.18$

and the joint scaling test are significant. In both cases, therefore, there is clear evidence of the inadequacy of the simple additive-dominance model.

10. Scales

A failure of the additive-dominance model to fit the data, such as we found with the last two examples considered in the previous Section, must imply that one (or more) of the assumptions on which the model is based is in fact invalid. Thus, for example, in constructing the model we have assumed that the genes show simple autosomal inheritance. If then some of them were sex-linked or if there were a maternal element in the determination of the character, or indeed if the pattern of inheritance departed from the simple autosomal in any other way, the model would not be appropriate and would be found to fail in its fit with an adequate body of observational results. This does not of course mean that biometrical analysis is impossible: it means only that a more appropriate model must be found and fitted to the data. The failure of the additive-dominance model in the examples of the last Section is, however, most unlikely to be due to invalidity of the assumption of simple autosomal inheritance. *Nicotiana rustica* and the tomato are both hermaphroditic plants and sex-linkage cannot therefore be involved. The reciprocal F_1's were alike in their expression of the character and this rules out a maternal element in its determination. There is no reason to postulate inviability of any of the genotypes included in the families raised, and the experiment was conducted in such a way as to minimize, if not entirely eliminate, the chance of selection disturbing the segregation of the genes.

These considerations point to the assumption of simple additivity of the d's and h's stemming from the various genes as the invalid part of the model. Again, as we shall see in Chapter 5, the model can be elaborated to accommodate non-independence of the effects of the different genes, although only at the expense of introducing further parameters. There is, however, one particular cause of non-independence whose effects can be resolved in a different way, so allowing the simple additive-dominance model to be retained and the complexity of introducing special parameters for the accommodation of the interactions among the genes to be avoided.

The additive-dominance model assumes that the genes involved are independent of each other in producing their effects; or in other words that the total effect of all the genes affecting the character (or at least the total effect of all such genes which affect the observations we are making) is the simple sum of their individual effects. Clearly this need not be so. Genes might, for example, act in a multiplicative fashion, that is their joint effect is the product, not the sum, of their individual actions, and such multiplicativity has in fact often been postulated. In such a case the simple model we have been using must fail when applied to an adequate body of data. But if two genes are acting in this way, their joint effect being $x_a x_b$, where x_a and x_b are their individual effects, and we replace the measurement of the phenotype by its logarithm we have $\log (x_a x_b) = \log x_a + \log x_b$. The multiplicative action has been removed and they now make their own independent contributions to the phenotype. So when in such a case we carry out the analysis in terms of the logarithms of our initial measurements, the assumption of independence is justified and the simple model will fit. Many other relations between genes and phenotype are obviously possible and each would suggest a suitable transformation of the scale on which the measurements of the phenotype are expressed to restore independence. To take but one more example, if the genes are additive in their effects on the linear dimensions of an organ while the character we are following is effectively an area it will reflect not the sum of the gene effects (as a linear character would) but the square of the sum. In respect of the area character, then, the model which assumes additivity will fail; but if we replace the direct observations by their square roots, so restoring to it a linear basis, the assumption of additive action of the genes would be valid and the model would fit these rescaled results. In other words where the assumption of independent action of the genes fails for this kind of reason, it is possible in principle to transform the data to a more appropriate scale, as by taking logs or square roots, or whatever else it may be, and to carry out the analysis successfully using the simple additive-dominance model on these transformed data.

The difficulty is, of course, that we cannot in general know how the genes affecting a character combine in producing their effects, or even whether in fact they all combine in the same way. So given that the model fails when applied to a set of data, we can only cast around for a transformation which removes, or at any rate substantially reduces, the non-independence. Sometimes the nature of the character may suggest a suitable transformation. Thus if a character effectively depends on the area of an organ, the square root transformation is an obvious one to try; but we must not be surprised if it fails, as we obviously cannot know that the genes combine additively in their effects on linear dimensions. In the same way the total weight of fruit yielded by say a tomato plant can be regarded as the product of number of fruits and their average weight. This is a multiplicative relation and suggests a log transformation; but again it does not follow that because these components of yield are related multiplicatively the genes affecting any one of the components combine in a similar way or that some genes do not affect both components simultaneously and so introduce a disturbance into the multiplicative relation.

Thus ultimately the only justification for any transformation that may be used is that it works; that whereas on the original data the model failed because of non-independence, once the data have been transformed the non-additivity vanishes, the simple model is adequate and there is no need to complicate the analysis or the interpretation of its results by introducing parameters to accommodate the non-additivity. Furthermore, because our test of the satisfactoriness of a transformation is empirical, by showing that it is successful in allowing analysis in terms of the simple model, we must be careful not to use its success as a justification for drawing theoretical conclusions concerning the physiology of gene action. At the same time, it is of course legitimate to test the agreement of any empirical scale with one expected theoretically from other considerations. This caution is reinforced when we consider that even where the genes are not all combining in the same way to produce their effects it may still be possible to find a scale on which their effects are independent on average, at least as far as the data under analysis go. In such a case it can give us little if any good information about the nature of gene action and interaction, and indeed this same transformation may fail when applied to a different cross involving different genes, as has in fact been observed to happen on many occasions in practice. Even, however, where this occurs, empirically the transformation has been justified since it has simplified the analysis of the body of data to which it was applicable and lent more precision and confidence to the predictive use of the results of that analysis.

We can see the value of a suitable transformation if we return to the example already considered on page 41, where the additive-dominanace model failed to fit the data on the weight per loculus of fruit in the cross between two tomato varieties (Table 8). Powers (1951) has published these data on both the original scale and on a logarithmic scale. We can, therefore, carry out the same tests on the log transformed data. These tests summarized in Table 9 provide clear evidence of the adequacy of

TABLE 9.

Analysis of weight per loculus of fruit in the tomato cross Danmark × Red Currant using the log transformed data (Powers, 1951). Compare with cross 1 in Table 8

Generation	Mean and its S.E. on logarithmic scale
P_1	0.9769 ± 0.026 61
P_2	−0.3643 ± 0.018 36
F_1	0.3346 ± 0.026 73
F_2	0.2726 ± 0.014 65
B_1	0.6357 ± 0.017 06
B_2	−0.0512 ± 0.014 67
Scaling tests	
A	−0.0401 ± 0.050 85
B	−0.0727 ± 0.043 73
C	−0.1914 ± 0.085 65
Joint	$\chi^2_{[3]} = 5.66$

the additive-dominance model on the new scale. In contrast the data on plant height in the cross between two *Nicotiana rustica* which were considered along with the tomato data (Table 8) could not be successfully transformed to a scale on which the simple model was adequate by taking logs, antilogs, squares or square roots of the original data. The further analysis of thcsc data is taken up in Chapter 5.

One last point remains to be made about scales of measurement. If we employ a transformation to remove interactions between non-allelic genes, as in the example we have just considered, we may, and indeed commonly will, change the apparent degree of dominance that the individual genes show, in other words change the value of the ratio h/d. This is well illustrated by the data in facet number in Bar-eyed female *Drosophila* quoted in Section 8. The comparisons among the facet numbers of B/B, B/+ and +/+ flies are shown in Table 10 using the direct counts of the facets, the logs of these counts and the square roots of them.

TABLE 10.

Effect of scalar transformation on the analysis of facet number in Bar-eyed *Drosophila*

Genotype	Mean facet number		
	Direct count	Log. transformation	Square-root
+/+ (wild type)	779.4	2.892	27.92
B/+	358.4	2.554	18.93
B/B	68.1	1.833	8.25
Components			
m	423.75	2.3625	18.085
d	355.65	0.5295	9.835
h	−65.35	0.1915	0.845
h/d	−0.184	0.362	0.086

As we have already seen, when the direct counts are used, h is negative and the Bar allele appears partially dominant to its wild-type alternative. If, however, we apply the log transformation, h becomes positive and h/d is larger than with the direct measure of facet number, so suggesting not only that wild-type is partially dominant to Bar but that the degree of dominance is larger as well as being in the opposite direction. But if we take the square root of facet number (which might be regarded as reasonable since the number of facets is essentially a measure of area), h/d is near to 0, so suggesting that dominance is in truth negligible.

Which of these scales we choose to use, and hence what direction and degree of dominance we choose to accept, is in this case a matter of taste, for with a gene difference of such large and unique effect by comparison with the residual variation in facet number, we have no test of whether any of the scales is preferable to the others in respect of reducing or removing interactions with other genes. If our aim is to simplify the representation of the effect of Bar, as far as possible, the square root transformation has the advantage of eliminating h and leaving us only with the need to use d in describing the relation between the three genotypes. At the same time, no matter which scale we use we can easily predict the mean facet number of an F_2, back-cross or any other type of family we care to consider, since in the absence of other segregating genes of comparable effect h and d give us a complete description of the genetic determination of the action of Bar. Furthermore, we should note that no matter which scale is used, we must conclude that dominance, if present, is small. Neither the log nor the square root transformation (nor for that matter, any other reasonable transformation) would show dominance as other than complete, i.e. $h = d$, if in fact B/+ had had the same number of facets as one or other of homozygotes, and neither transformation would have failed to reveal over-dominance, i.e. $h > d$, if the facet number of B/+ had fallen outside the range determined by B/B and +/+.

As has been emphasized, the justification for using a transformed scale is not theoretical but empirical, in that it removes or so reduces non-independence of the gene effects as to permit the use of the additive-dominance model with the simpler analysis and more confident prediction to which it leads. Furthermore the estimates of the genetical parameters d and h, obtained when the additive-dominance model can be employed, are unconditional in that they are not subject to adjustment by the interaction parameters which non-additivity introduces and are constant over the range of variation under consideration. For these reasons, while we must recognize that it is not always possible to find a transformation which in effect removes non-additivity when this is present in the direct measurements, the search for such a transformation is always well worth-while.

11. Components of variation: F_2 and back-crosses

So far we have been considering the constitution of family means in terms of the additive-dominance model and the way in which observational data can be analysed so as to yield not only estimates of the genetical parameters $[d]$ and $[h]$, in terms of which the values of the means can be interpreted, but also a test of whether the model fits the data in the sense of providing an adequate framework for the understanding of the observations. We must now leave these first degree statistics, the means, and turn to consider the second degree statistics, the variances and covariances that can be calculated from the families raised in genetical experiments, the genetical parameters in terms of which these statistics can be analysed and the test of whether the simple model provides an adequate basis for understanding them.

Now the variation in each of the true breeding parent lines, P_1 and P_2, must be exclusively non-heritable, for all the individuals within one line will be of the same genotype, apart from the effects of mutation which, although detectable in suitable experiments, are in general so small as to be safely neglected. Similarly all the individuals in the F_1 between two such parent lines will have the same genotype although they will be heterozygous and not homozygous like their parents. Again all the variation will be non-heritable within the F_1 family as it was in the parents. The variances of the measurements of the character in both parents and F_1 will thus provide estimates of the non-heritable variation and of its contribution to the variances of later generations in which, because of segregation of the genic differences between P_1 and P_2, heritable variation will also be present.

Considering first the F_2, in the absence of disturbing elements such as differential fertilization or viability, its constitution in respect of any gene pair A-a by which P_1 and P_2 differed, will be $\frac{1}{4}$AA, $\frac{1}{2}$Aa and $\frac{1}{4}$aa. This gene pair will add increments of d_a, h_a and $-d_a$ to the expression of the character in individuals of the three genotypes and, as we have already seen (Table 5) the contribution of A-a to the deviation of the F_2 mean from m, the mid-parent, will be $\frac{1}{2}h_a$. The contribution of A-a to the sum of squares of deviation from the mid-parent will be

$$\tfrac{1}{4}d_a^2 + \tfrac{1}{2}h_a^2 + \tfrac{1}{4}(-d_a)^2 = \tfrac{1}{2}d_a^2 + \tfrac{1}{2}h_a^2$$

and its contribution to the sum of squares from the F_2 mean then becomes

$$\tfrac{1}{2}d_a^2 + \tfrac{1}{2}h_a^2 - (\tfrac{1}{2}h_a)^2 = \tfrac{1}{2}d_a^2 + \tfrac{1}{4}h_a^2$$

the term correcting for the departure of the mean from the mid-parent being the square of the mean itself since we are using the proportionate frequencies of the three genotypes and these sum to unity. For the same reason the contribution of A-a to the mean square measuring the heritable variation, is the same as its contribution to the sum of squares, namely $\frac{1}{2}d_a^2 + \frac{1}{4}h_a^2$.

Assuming that non-allelic genes make independent contributions to it, the heritable variance produced by all the genes segregating in the F_2 will be the sum of their individual contributions. It thus becomes $\frac{1}{2}S(d^2) + \frac{1}{4}S(h^2) = \frac{1}{2}D + \frac{1}{4}H$ where we define $D = S(d^2)$ and $H = S(h^2)$. Thus the heritable variance comprises two parts, the D component, depending on the d's which measure the departure of homozygotes from the mid-parent and the H component which depends on the h's measuring the departures of heterozygotes from the mid-parent. The D variation can in principle be fixed by the selection of homozygous lines and so may be referred to as fixable variation. The H variation depends on the properties of heterozygotes and is therefore unfixable. H may also be described as the dominance component of variation since when dominance is absent at all loci, all h's $= 0$ and $H = 0$. Similarly if dominance is complete at all loci, all $h = \pm d$ and $H = D$, while with overdominance at all loci all $h > \pm d$ and $H > D$. Now since $D = S(d^2)$ and $H = S(h^2)$ both are quadratic quantities. By contrast therefore with $[d]$ and $[h]$, the values of D and H will be uninfluenced by the distribution between the parent lines of the alleles at the various loci and by the direction of dominance as reflected in the sign of h. Thus if we care to assume that h and d are constant in magnitude (although in the case of h not necessarily in sign) for all the genes segregating in the cross, $\sqrt{(H/D)} = h/d$ provides a direct estimate of the degree of dominance free of the disturbances which we had occasion to note when we were discussing the ratio $[h]/[d]$. If h and d are not constant in magnitude $\sqrt{(H/D)}$ provides an estimate of the average dominance of the genes.

Before leaving the variance of F_2 we should note that it must of course also include a non-heritable component which, provided the heritable and non-heritable components are independent of one another (i.e. provided that the phenotypes given by all the genotypes are subject to the same variation from non-heritable causes), can be denoted by a separate term E. Thus the variance of F_2 may be expressed as

$$V_{1F2} = \tfrac{1}{2}D + \tfrac{1}{4}H + E.$$

The reason for using V_{1F2} rather than the simple V_{F2} to denote this variance will appear later (page 52).

Proceeding from F_2 to the back-crosses we note that in respect of A-a the back-cross to the larger parent, P_1, will comprise $\frac{1}{2}$AA and $\frac{1}{2}$Aa individuals and that to the smaller parent, P_2, $\frac{1}{2}$Aa and $\frac{1}{2}$aa individuals. Then, as we have already seen, $\bar{B}_1 = \frac{1}{2}d_a + \frac{1}{2}h_a$ and $\bar{B}_2 = \frac{1}{2}h_a - \frac{1}{2}d_a$. The contributions of A-a to the variances of the two back-crosses will thus be,

$$\tfrac{1}{2}d_a^2 + \tfrac{1}{2}h_a^2 - [\tfrac{1}{2}(d_a + h_a)]^2 = \tfrac{1}{4}(d_a - h_a)^2 \text{ to } V_{B1}$$

and similarly $\frac{1}{4}(d_a + h_a)^2$ to V_{B2}. Then assuming independence of the contributions of the different genes, the heritable portions of the back-cross variances become $\frac{1}{4}\mathrm{S}(d-h)^2$ and $\frac{1}{4}\mathrm{S}(d+h)^2$ respectively. Clearly d and h do not make independent contributions and we must introduce a further component of variation, $F = \mathrm{S}(dh)$, to give the expressions

$$V_{B1} = \tfrac{1}{4}D - \tfrac{1}{2}F + \tfrac{1}{4}H + E \quad \text{and} \quad V_{B2} = \tfrac{1}{4}D + \tfrac{1}{2}F + \tfrac{1}{4}H + E,$$

E representing the non-heritable variation as before. We may note, however, that if we add the two variances

$$V_{B1} + V_{B2} = \tfrac{1}{2}D + \tfrac{1}{2}H + 2E$$

and again we have an expression to which d and h make independent contributions. Similarly, if we take the difference of the two variances

$$V_{B1} - V_{B2} = F = \mathrm{S}(dh).$$

Now F is a linear function of the h's and so, like h, can take sign: it is in fact a weighted sum of the h's, the weights being the corresponding d's. Where F is positive the genes from the larger parent, P_1, show a preponderance of dominance over their alleles from P_2, and where F is negative the genes from the smaller parent P_2, show the preponderance of dominance. It will be observed too that because of F the back-cross to the parent with the preponderance of dominance gives the smaller variance.

If we assume that all k gene pairs by which P_1 and P_2 differ have equal d's and equal h's, $D = \mathrm{S}(d^2) = kd^2$, $H = \mathrm{S}(h^2) = kh^2$ and $F = \mathrm{S}(dh) = kdh$. Then $\sqrt{(DH)} = \sqrt{(kd^2 . kh^2)} = kdh = F$, provided the h's are all of the same sign. But if the h's vary in their sign, some being + and others −, $F < \sqrt{(DH)}$. Exactly the same conclusions are arrived at even when we do not have equal d's and h's providing that the dominance ratio h/d, is the same for all k loci. We have, therefore, in principle a test of consistency in the sign of the h's.

When analysing the components of variation the simple additive-dominance model assumes that the various gene pairs contribute independently to the variances and covariances just as we saw that it did when analysing the components of means. In addition, however we now have the further assumption that the contribution to the variation made by non-heritable agencies is independent of that made by the genes, or to put it in other words that there is no interaction of genotype and environment. This is by no means always a valid assumption, for we not uncommonly find different genotypes to be subject to different types of non-heritable variation. Sometimes the differences can be removed, or at least greatly reduced by a transformation of the scale.

Commonly, however, we find that an F_1 between two inbred lines of a naturally outbreeding species, while showing an intermediate mean expression of a character shows a variance lower than those of both parents. No reasonable transformation of the scale will remove such differences. Two courses are then open. A simple, if somewhat crude, allowance for the differences can be made by taking the average of the parental and F_1 variances as the direct estimate of E; and this can be refined by an appropriate weighting of the contributions the parents and F_1 make to the average, for example, by taking $\frac{1}{4}V_{P1} + \frac{1}{4}V_{P2} + \frac{1}{2}V_{F1}$ (where V_{P1} is the variance of parent 1 etc.) as a direct estimate of the E component in V_{IF2}, and in the summed variances of the back-crosses, $V_{B1} + V_{B2}$. Difficulties arise when we move on to later generations, since the corresponding weighting should change, as for example in F_3 where E in the overall variance should be found as $\frac{3}{8}V_{P1} + \frac{3}{8}V_{P2} + \frac{1}{4}V_{F1}$ since only $\frac{1}{4}$ of the individuals in F_3 are heterozygous at any locus by comparison with $\frac{1}{2}$ in F_2. Probably when making this simple correction for differences in the non-heritable variation among parents and F_1, putting $E = \frac{1}{4}V_{P1} + \frac{1}{4}V_{P2} + \frac{1}{2}V_{F1}$ is as useful a weighting as any, and well within the limits of error of such a crude, empirical correction.

The second course open to us is to expand the model and introduce into it appropriate parameters to represent the genotype X environment interaction in the way we shall see in Chapter 6. Such an expanded model, however, necessarily requires more data to permit the estimation of the greater number of parameters it entails and the testing of its goodness of fit. The use of a suitable transformation or a simple, if necessarily approximate, correction is always worth considering if the simple additive-dominance model can thereby be made to fit satisfactorily.

12. Generations derived from F_2

Further generations can be derived from the F_2 and the back-crosses, and the structures of their variances expressed in terms of D, H and F. Those from the back-crosses will not be considered here: they are dealt with by Mather and Jinks (1971).* In respect of the gene for A-a the overall composition of an F_3 generation, derived by selfing the individuals of F_2 will

* Since this reference will be in frequent use, it will hereafter be abbreviated to M and J.

be $\frac{3}{8}$AA; $\frac{1}{4}$Aa; $\frac{3}{8}$aa giving a mean of $\frac{1}{4}h_a$. The contribution of A-a to the variance V_{F3} will thus be $\frac{3}{8}d_a^2 + \frac{1}{4}h_a^2 + \frac{3}{8}(-d_a)^2 - (\frac{1}{4}h_a)^2 = \frac{3}{4}d_a^2 + \frac{3}{16}h_a^2$. This overall variance can, however, be broken down into two parts: the variance of the means of the F_3 families, V_{1F3}, round the overall mean of the F_3 generation, and the mean variance of the F_3 families, V_{2F3}, each calculated round its own mean but averaged over all families. The variance of the F_3 means is like the variance of F_2 in that its heritable portion reflects the genetical differences produced by segregation at gametogenesis of the F_1. These are therefore described as first rank variances, denoted by the subscript 1. The variances within the F_3 families themselves, however, reflect the segregation at gametogenesis of the F_2 individuals and the mean variance of the F_3 families is thus of the second rank, denoted by the subscript 2. As we shall see later, rank is of special significance in relation to the effects of linkage on the components of variation.

In respect of A-a, the F_3 families will be of three kinds derived respectively by selfing AA, Aa and aa individuals of the F_2. The families from homozygous F_2 individuals will be like P_1 and P_2 in the contribution A-a makes to their means and variances and the families from Aa individuals of F_2 will be like the F_2 itself in the contribution to mean and variance, thus

F_2 individuals		AA	Aa	aa
Frequency in F_2		$\frac{1}{4}$	$\frac{1}{2}$	$\frac{1}{4}$
F_3 family	mean	d_a	$\frac{1}{2}h_a$	$-d_a$
	variance	0	$\frac{1}{2}d_a^2 + \frac{1}{4}h_a^2$	0

The contribution to the variance of F_3 means, V_{1F3}, will thus be $\frac{1}{4}d_a^2 + \frac{1}{2}(\frac{1}{2}h_a)^2 + \frac{1}{4}(-d_a)^2 - (\frac{1}{4}h_a)^2$ the last term being the correction for the overall mean of $\frac{1}{4}h_a$. This reduces to $\frac{1}{2}d_a^2 + \frac{1}{16}h_a^2$, which summing over all the genes by which P_1 and P_2 differ gives $\frac{1}{2}D + \frac{1}{16}H$ as the heritable portion of V_{1F3}. The contribution of A-a to the mean variance, V_{2F3}, will be $\frac{1}{4}(0) + \frac{1}{2}(\frac{1}{2}d_a^2 + \frac{1}{4}h_a^2) + \frac{1}{4}(0) = \frac{1}{4}d_a^2 + \frac{1}{8}h_a^2$ which on summing over all gene differences gives $\frac{1}{4}D + \frac{1}{8}H$ as the heritable portion of the mean variance.

Both these variances will of course also contain a non-heritable component, E, but these E components will not in general be equal. In the first place the effect of those non-heritable agencies that cause differences among the members of a family will be less on the mean of the family than on its individual members. Indeed in respect of this part of the non-heritable variation $E_2 = \frac{1}{n}E_1$, where E_2 is the variation of the means of families comprising n individuals each and E_1 is the variation within the families. But where each family is raised in its own plot in the case of plants, or in its own cage or culture container in the case of animals, we must expect greater non-heritable differences between individuals from different families, i.e. coming from different plots or containers, than between individuals from the same family, i.e. from the same plot or container. Thus, unless special experimental designs are used to avoid this situation, we must expect $E_2 > \frac{1}{n}E_1$ and in extreme cases E_2 may even be greater than E_1 itself. If we write E_w for the non-heritable variation within families and E_b for the additional non-heritable variation between families, we can put $E_2 = E_b + \frac{1}{n}E_w$, and, of course, $E_1 = E_w$.

There is another point to be noted about the variance of family means. Each mean will be subject to sampling variation arising from the variation within the family, and this will be additional to the innate variation between the family means themselves, arising from genetical or indeed any other differences between the means as such. The component of sampling variation in V_{1F3} will be $\frac{1}{n}V_{2F3}$ where each family includes n individuals, or, if the numbers vary from one family to another, where n is the harmonic mean of these numbers. $\frac{1}{n}V_{2F3}$ will of course include the item $\frac{1}{n}E_w$, which is the contribution of sampling variation in respect of non-heritable variation within families to non-heritable variation between their means. We can thus write

$$V_{1F3} = \tfrac{1}{2}D + \tfrac{1}{16}H + E_b + \tfrac{1}{n}V_{2F3}$$
$$V_{2F3} = \tfrac{1}{4}D + \tfrac{1}{8}H + E_w.$$

In addition to these two variances we can also find the covariance, W_{1F23}, between the phenotype of the F_2 parent and the mean of the F_3 family to which it gives rise. This covariance will of course be of the first rank. In respect of A-a, an AA F_2 individual will have a phenotype of d_a and will give rise to a progeny of mean d_a. Similarly an aa F_2 individual will have a phenotype $-d_a$ and the mean of its progeny will be $-d_a$; but an Aa individual in F_2 will have a phenotype h_a itself while the mean of its progeny will only be $\frac{1}{2}h_a$. The contribution of A-a to the covariance will thus be $\frac{1}{4}(d_a)^2 + \frac{1}{2}(h_a.\frac{1}{2}h_a) + \frac{1}{4}(-d_a)^2 - \frac{1}{2}h_a.\frac{1}{4}h_a$, the correction term being the product of the F_2 and overall F_3 means. This reduces to $\frac{1}{2}d_a^2 + \frac{1}{8}h_a^2$ and, summing over all the relevant genes, gives $W_{1F23} = \frac{1}{2}D + \frac{1}{8}H$.

There will be no E component in the covariance provided that the non-heritable agencies affecting the progeny are uncorrelated with those affecting the parents. This lack of correlation can be achieved, and an E component avoided, by independent randomization of parents and offspring in the experiment, so that they do not share a common family environment. Such independent randomization is a standard practice in experimental plant breeding; but it is difficult to achieve with higher animals because of the essential period of maternal care for the young offspring, with the consequence that the covariance must be expected to contain an E component in such cases.

We can extend the calculations to the F_4 generation, where there are three variances and two covariances. The first variance, V_{1F4}, is that between the means of the groups of F_4 families, where the members of each group trace back through a single F_3 family to a single F_2 individual, and it is therefore of rank 1. There will be a corresponding covariance, W_{1F34}, between the means of the F_3 families and the means of the F_4 groups. The second variance, V_{2F3}, is the variance of F_4 family means within the groups taken round the group means but averaged over groups. It will be of rank 2, and will have a corresponding covariance, W_{2F34}, between F_3 individuals and the mean of the F_4 families to which they give rise, calculated within groups but averaged over groups. Finally there will be the mean variance of families averaged over all the F_4 families, which will be of rank 3 since it reflects differences springing from gametogenesis in the F_3 individuals. Provided that E_b is no greater between families from different groups than between those of the same group, and making allowance for the appropriate sampling variation of family and group means, with n individual in each family and n' families in each group, it can be shown that

$$\begin{aligned}
V_{1F4} &= \tfrac{1}{2}D + \tfrac{1}{64}H + \tfrac{1}{n'}V_{2F4} \\
V_{2F4} &= \tfrac{1}{4}D + \tfrac{1}{32}H + E_b + \tfrac{1}{n}V_{3F4} \\
V_{3F4} &= \tfrac{1}{8}D + \tfrac{1}{16}H + E_w \\
W_{1F34} &= \tfrac{1}{2}D + \tfrac{1}{32}H \\
W_{2F34} &= \tfrac{1}{4}D + \tfrac{1}{16}H.
\end{aligned}$$

We can proceed in the same way to F_5, where there will be four variances and three covariances, and indeed to any later F generation that we wish.

In addition to F_3's other types of family can be raised from F_2 parents. The F_2 individuals may for example be mated together in random pairs to give families of the type that Mather (1949) has called BIPS (for biparental progenies of the third generation). Such random mating of the

F_2 individuals will obviously give a third generation which (linkage apart) has overall the same constitution as the F_2 itself, and which will thus have an overall mean of $\frac{1}{2}[h]$ and an overall variance of $\frac{1}{2}D + \frac{1}{4}H + E$. As with the F_3, however, we can divide this overall variance into two parts, the variance of the family means (V_{1S3}) and the mean variance of the families (V_{2S3}), the subscript S indicating sib-mating and so allowing extension of the nomenclature to fourth and later generations raised by random sib-mating within families. In respect of any gene pair, A-a, there are six types of mating among the F_2 individuals. These, together with their frequencies where mating is at random, their means and their variances, in respect of A-a, are shown in Table 11. It is not difficult to see

TABLE 11.

Biparental progenies from random matings among the individuals of an F_2

Mating	Frequency	Progeny	
		Mean	Variance
AA × AA	$\frac{1}{16}$	d	0
AA × Aa	$\frac{1}{4}$	$\frac{1}{2}(d+h)$	$\frac{1}{4}(d-h)^2$
AA × aa	$\frac{1}{8}$	h	0
Aa × Aa	$\frac{1}{4}$	$\frac{1}{2}h$	$\frac{1}{2}d^2+\frac{1}{4}h^2$
Aa × aa	$\frac{1}{4}$	$\frac{1}{2}(h-d)$	$\frac{1}{4}(d+h)^2$
aa × aa	$\frac{1}{16}$	$-d$	0
Overall mean $\frac{1}{2}h$			

from this table that the contribution of A-a to the variance of family means (V_{1S3}) will be

$$\tfrac{1}{16}d_a^2 + \tfrac{1}{4}[\tfrac{1}{2}(d_a + h_a)]^2 \ldots \tfrac{1}{16}(-d_a)^2 - (\tfrac{1}{2}h_a)^2 = \tfrac{1}{4}d_a^2 + \tfrac{1}{16}h_a^2$$

where the term $-(\frac{1}{2}h_a)^2$ is the correction for the deviation of the overall mean of the generation from the mid-parent m. Similarly the contribution of A-a to the mean variance of the families (V_{2S3}) will be $\frac{1}{16}(0) + \frac{1}{4}\cdot\frac{1}{4}(d_a - h_a)^2 \ldots + \frac{1}{16}(0) = \frac{1}{4}d_a^2 + \frac{3}{16}h_a^2$. Then summing over all the relevant genes, adding the non-heritable component of variation and also the item for sampling variation in V_{1S3}, we find

$$V_{1S3} = \tfrac{1}{4}D + \tfrac{1}{16}H + E_b + \tfrac{1}{n}V_{2S3}$$

$$V_{2S3} = \tfrac{1}{4}D + \tfrac{3}{16}H + E_w$$

to which may be added

$$W_{1S23} = \tfrac{1}{4}D$$

for the covariance of the family means with the phenotypes of their F_2 parents. We can proceed in the same way to S_4, the fourth generation raised by random sib-mating inside the F_3 families where just as with F_4, there will be three variances and two covariances, and indeed to later generations (see M and J).

These results are collected together in Table 12. A fuller compilation is given by M and J (Table 44) which includes also the constitution of variances and covariances from later generations derived from the back-crosses.

TABLE 12.

Components of variation in F_2 and its derivatives

Statistic	D	H	E_w	E_b	Sampling variation
V_{1F2}	$\frac{1}{2}$	$\frac{1}{4}$	1	0	0
V_{1F3}	$\frac{1}{2}$	$\frac{1}{16}$	0	1	$\frac{1}{n} V_{2F3}$
V_{2F3}	$\frac{1}{4}$	$\frac{1}{8}$	1	0	0
W_{1F23}	$\frac{1}{2}$	$\frac{1}{8}$	0	0	0
V_{1F4}	$\frac{1}{2}$	$\frac{1}{64}$	0	0	$\frac{1}{n'} V_{2F4}$
V_{2F4}	$\frac{1}{4}$	$\frac{1}{32}$	0	1	$\frac{1}{n} V_{3F4}$
V_{3F4}	$\frac{1}{8}$	$\frac{1}{16}$	1	0	0
W_{1F34}	$\frac{1}{2}$	$\frac{1}{32}$	0	0	0
W_{2F34}	$\frac{1}{4}$	$\frac{1}{16}$	0	0	0
V_{1S3}	$\frac{1}{4}$	$\frac{1}{16}$	0	1	$\frac{1}{n} V_{2S3}$
V_{2S3}	$\frac{1}{4}$	$\frac{3}{16}$	1	0	0
W_{1S23}	$\frac{1}{4}$	0	0	0	0
V_{1S4}	$\frac{1}{4}$	$\frac{3}{128}$	0	0	$\frac{1}{n'} V_{2S4}$
V_{2S4}	$\frac{1}{8}$	$\frac{5}{128}$	0	1	$\frac{1}{n} V_{3S4}$
V_{3S4}	$\frac{1}{4}$	$\frac{11}{64}$	1	0	0
W_{1S34}	$\frac{1}{4}$	$\frac{1}{32}$	0	0	0
W_{2S34}	$\frac{1}{8}$	$\frac{1}{32}$	0	0	0

13. The balance sheet of genetic variability

Like energy, genetic variability is conserved inside a closed system. Crossing, segregation and recombination, may redistribute it among the various states in which it can exist, but in the absence of mutation, random change and selection its total quantity remains unchanged (see Mather, 1973 for a fuller discussion of the theory of variability). One aspect of this conservation of variability is revealed by the heritable variances we have been discussing.

The heritable portion of the phenotypic differences between homozygotes is D- type variation. Heterozygotes contribute to the phenotypic differences in two ways. They may contribute directly to the phenotypic differences among the individuals of a family or generation; but their contribution may also appear in part as the departure of the generation mean from the mid-parent, which as we have seen depends on $[h]$. Now D and H are both quadratic quantities, in terms of d and h, but $[h]$ on the other hand is linear. The coefficient of $[h]$ in the departure of the mean from the mid-parent must thus be squared if it is to be comparable to the coefficients of D and H. The heritable variation expressed by the phenotypes of a generation may thus be expressed as $xD + yH + z[h]^2$ and in the absence of complicating circumstances, x, y and z must sum to unity.

In the F_1, $x = y = 0$ and $z = 1$ since the mean is $[h]$; but in the F_2 to which it gives rise $x = \frac{1}{2}$, $y = \frac{1}{4}$ and with the mean at $\frac{1}{2}[h]$, $z = \frac{1}{2}^2 = \frac{1}{4}$ so once again giving $x + y + z = \frac{1}{2} + \frac{1}{4} + \frac{1}{4} = 1$. The F_3 has an overall mean of $\frac{1}{4}[h]$ so giving $z = \frac{1}{4}^2 = \frac{1}{16}$. There are two variances whose heritable components are to be taken into account in the F_3. These are $V_{1F3} = \frac{1}{2}D + \frac{1}{16}H$ and $V_{2F3} = \frac{1}{4}D + \frac{1}{8}H$, sampling variation being left out of account as any differences it produces are random changes. Thus taken together these two variances contribute $\frac{3}{4}D + \frac{3}{16}H$ and $x = \frac{3}{4}$, $y = \frac{3}{16}$ while as we have seen $z = \frac{1}{16}$ so completing the tally and giving $x + y + z = 1$. The same applies to F_4 (see Table 13) and indeed to F_5 or any later generation. In the biparental progenies of the third generation the heritable components of the two variances are $V_{1S3} = \frac{1}{4}D + \frac{1}{16}H$ and $V_{2S3} = \frac{1}{4}D + \frac{3}{16}H$ while the mean is $\frac{1}{2}[h]$. So $x = \frac{1}{4} + \frac{1}{4} = \frac{1}{2}$, $y = \frac{1}{16} + \frac{3}{16} = \frac{1}{4}$, $z = (\frac{1}{2})^2 = \frac{1}{4}$ giving once again $x + y + z = 1$, and the same can be shown to apply to S_4 the fourth generation, and indeed to S_5 etc. raised by continued sib-mating (see M and J).

It will be observed that the coefficient of D in the successive F generations follows, as indeed it must, the series $1 - \frac{1}{2}^{n-1}$ which gives the proportion of individuals homozygous in the nth generation for the alleles

TABLE 13.

The balance sheet of variability

Generation		Coefficient of			
		D	H	$[d]^2$	$[h]^2$
Parents		0	0	$1^2=1$	0
F_1		0	0	0	$1^2=1$
F_2		$\frac{1}{2}$	$\frac{1}{4}$	0	$(\frac{1}{2})^2=\frac{1}{4}$
F_3	V_{1F3}	$\frac{1}{2}$	$\frac{1}{16}$		
	V_{2F3}	$\frac{1}{4}$	$\frac{1}{8}$		
	Total	$\frac{3}{4}$	$\frac{3}{16}$	0	$(\frac{1}{4})^2=\frac{1}{16}$
F_4	V_{1F4}	$\frac{1}{2}$	$\frac{1}{64}$		
	V_{2F4}	$\frac{1}{4}$	$\frac{1}{32}$		
	V_{3F4}	$\frac{1}{8}$	$\frac{1}{16}$		
	Total	$\frac{7}{8}$	$\frac{7}{64}$	0	$(\frac{1}{8})^2=\frac{1}{64}$
S_3	V_{1S3}	$\frac{1}{4}$	$\frac{1}{16}$		
	V_{2S3}	$\frac{1}{4}$	$\frac{3}{16}$		
	Total	$\frac{1}{2}$	$\frac{1}{4}$	0	$(\frac{1}{2})^2=\frac{1}{4}$
S_4	V_{1S4}	$\frac{1}{4}$	$\frac{3}{128}$		
	V_{2S4}	$\frac{1}{8}$	$\frac{5}{128}$		
	V_{3S4}	$\frac{1}{4}$	$\frac{11}{64}$		
	Total	$\frac{5}{8}$	$\frac{15}{64}$	0	$(\frac{3}{8})^2=\frac{9}{64}$
Back-crosses	$V_{\bar{B}}$	$\frac{1}{4}$	$\frac{1}{4}$	0	
	V_{B}	0	0	$\frac{1}{4}$	
	Total	$\frac{1}{4}$	$\frac{1}{4}$	$\frac{1}{4}$	$(\frac{1}{2})^2=\frac{1}{4}$

at a locus at which the parents differed. Similarly the sum of the coefficient of H and the squared coefficient of $[h]^2$ follows the series $\frac{1}{2}^{n-1}$, since the proportion of heterozygotes at such a locus is halved in each generation under selfing. In the same way the coefficients of D, H and $[h]^2$ in S_3, S_4 etc. are related to the Fibonacci series which gives the fall in the proportion of heterozygotes under continued sib-mating.

The same principle of conservation of variability applies to the joint back-crosses although with the introduction of a fourth component. The heritable portion of the mean variance of the two back-crosses is $\overline{V}_B = \frac{1}{2}(V_{B1} + V_{B2}) = \frac{1}{4}D + \frac{1}{4}H$. The means of the back-crosses are $\overline{B}_1 = \frac{1}{2}([d] + [h])$ and $\overline{B}_2 = \frac{1}{2}([h] - [d])$ the overall mean of the two taken together being $\frac{1}{2}[h]$. The heritable variance of the back-cross means is thus

$$V_{\overline{B}} = \tfrac{1}{2}\,[\tfrac{1}{2}([d] + [h])]^2 + \tfrac{1}{2}\,[\tfrac{1}{2}([h] - [d])]^2 - (\tfrac{1}{2}[h])^2 = \tfrac{1}{4}[d]^2.$$

The departure of the overall mean from the mid-parent accounts for $(\frac{1}{2}[h])^2 = \frac{1}{4}[h]^2$ of the variability, and the coefficients of D, H, $[h]^2$ and the new component $[d]^2$ thus sum to unity (Table 13). Once this fourth component of variability is recognized we can complete the picture by noting that in the parental generation, $\overline{P}_1 = [d]$ and $\overline{P}_2 = -[d]$, giving a total of $[d]^2$ for the variability represented by the difference between the means of these two true-breeding lines from whose cross all the later generations are descended.

In conclusion we should note that D, H, $[d]^2$ and $[h]^2$ are different components of variability with different properties. Their coefficients sum to unity because all the variability must be acounted for, but each of them has its own special relation to the expression of variability among the phenotypes. Thus H and $[h]^2$ depend on dominance while D and $[d]^2$ do not. The dominance properties of the genes express themselves in different ways in $[h]^2$ than in H: dominance in opposing directions tends to balance out in $[h]^2$ but not in H. Furthermore $[h]^2 \neq H$ apart from the trivial case where only one gene difference is involved, for even where all the gene pairs show dominance in the same direction $[h]^2$ will exceed H by a factor which depends on how many gene pairs are involved and by how much the individual h's vary from one to another. In the same way $[d]^2$ will reflect the distribution of the genes between the parents whereas D will not: thus D will be the same in the cross AABB × aabb as in AAbb × aaBB, whereas $[d]^2$ will not. And where all the increasing alleles are associated in one parent, AA BB CC . . . . . , and all the decreasing alleles in the other, aa bb cc . . . , $[d]^2$ will exceed D by a factor depending on the number of gene pairs involved and on the extent to which the individual d's vary from one to another. We shall have occasion again to touch on these relationships in a later section.

14. Partitioning the variation

The D, H and E components of variation differ in the relative contributions they make to the variances and covariances in the various generations and types of family we can raise from a cross between two true-breeding parental lines. We can therefore obtain estimates of these components by suitable comparisons among the various second degree statistics. This is seen at its simplest if we turn again to the example described on page 37 of plant height in the P_1, P_2, F_1, F_2, B_1 and B_2 families raised from the 22 X 73 cross of *Nicotiana rustica.* The earlier analysis showed that a simple additive-dominance model satisfactorily accounted for the means of these generations. Now we shall consider the variances of these same families and obtain estimates of D, H, F and E. These variances are set out in Table 14. Although we have six variances three of them (V_{P1},

TABLE 14.

Variances within families for plant height in the cross between varieties 22 and 73 of *Nicotiana rustica* (corresponding with the means in Table 6)

Family	Variance	Expectation
P_1	20.6684	E_w
P_2	29.0500	E_w
F_1	57.4260	E_w
F_2	77.6533	$\frac{1}{2}D + \frac{1}{4}H + E_w$
B_1	59.5288	$\frac{1}{4}D + \frac{1}{4}H - \frac{1}{2}F + E_w$
B_2	66.1747	$\frac{1}{4}D + \frac{1}{4}H + \frac{1}{2}F + E_w$
Components		
D	59.2062	
H	27.6304	
F	6.6459	
E_w	41.1426	(found as $\frac{1}{4} V_{P1} + \frac{1}{4} V_{P2} + \frac{1}{2} V_{F1}$)
$\sqrt{\frac{H}{D}}$	0.6831	(Dominance ratio)
$\frac{F}{\sqrt{(DH)}}$	0.1643	

V_{P2} and V_{F1}) are all estimates of E_w. Two of these, from the two parental families, do not differ from one another, but they do differ from the F_1 estimate which is significantly larger. We must therefore combine them in the way described on p. 51, to give

$$E_w = \tfrac{1}{4}(V_{P1} + V_{P2} + 2V_{F1}) = 41.1426.$$

The combined estimate of E_w together with the remaining three variances leave us with four equations for estimating the four components D, H, F, and E_w. So only a perfect fit solution is possible, the equations being

$$D = 4V_{1F2} - 2(V_{B1} + V_{B2}) = 59.2062$$
$$H = 4(V_{B1} + V_{B2} - V_{1F2} - E_w) = 27.6304$$
$$F = V_{B2} - V_{B1} = 6.6459.$$

These estimates are tabulated in Table 14. Finally we can estimate the dominance ratio as $\sqrt{H/D}) = 0.6831$ which agrees with the relatively high level of dominance suggested by the analysis of the means. The relatively low value for $F/\sqrt{(D \cdot H)}$ provides little evidence that the dominance deviations at different loci are particularly consistent in sign or magnitude. Having only four equations for the estimation of four parameters we must obtain a perfect fit solution to them, and we can neither calculate the standard deviation of the estimates of D, H, E and F, nor indeed can we test the goodness of fit of the additive-dominance model as a whole. To do so requires a more comprehensive experiment such as that described and analysed by Hayman (1960), which is also discussed by M and J.

Hayman's experiment was again initiated by a cross between two true-breeding lines of *Nicotiana rustica*, although it was not the same cross as the one we have just been considering. The two parents were crossed reciprocally to give the two reciprocal F_1's from each of which an F_2, F_3 and F_4 were raised. The F_3 consisted of 10 families from each reciprocal, i.e. 20 F_3's in all, and the F_4 of 50 families from each reciprocal, the 100 families thus involved being obtained by selfing 5 plants from each of 20 F_3 families. Back-crosses were not included in the experiment. The character we shall be considering is plant height measured in inches. The plants were grown in two blocks, the plots within the blocks each comprising five plants. Each of the F_3 and F_4 families occupied one plot in each block, but each parent, F_1 and F_2 was present as five plots in each of the two blocks. There is internal evidence from Hayman's account of the experiment that some F_4 plants, and it would appear seven F_4 families, failed in the experiment or were excluded for other reasons. V_{1F2}, V_{2F3} and V_{3F4} were obtained from the variances within plots, round the plot means, and so include E_w as their non-heritable component. V_{1F3}, V_{1F4} and V_{2F4} were found as variances between the relevant plot means, taken round the block means, and so include E_b as well as the sampling variation stemming from V_{2F3}, V_{2F4} and V_{3F4} respectively. Since each plot included five plants, $n = 5$ and in F_4 each group included five families so giving $n' = 5$ also. Thus allowing for sampling variation (see pp. 53–4)

$$\begin{aligned} V_{1F3} &= \tfrac{1}{2}D + \tfrac{1}{16}H + E_b + \tfrac{1}{5}V_{2F3} \\ &= \tfrac{1}{2}D + \tfrac{1}{16}H + E_b + \tfrac{1}{5}(\tfrac{1}{4}D + \tfrac{1}{8}H + E_w) \end{aligned}$$

and similarly

$$\begin{aligned} V_{2F4} &= \tfrac{1}{4}D + \tfrac{1}{32}H + E_b + \tfrac{1}{5}V_{3F4} \\ &= \tfrac{1}{4}D + \tfrac{1}{32}H + E_b + \tfrac{1}{5}(\tfrac{1}{8}D + \tfrac{1}{16}H + E_w). \end{aligned}$$

Since $n' = 5$

$$\begin{aligned} V_{1F4} &= \tfrac{1}{2}D + \tfrac{1}{64}H + \tfrac{1}{n'}V_{2F4} \\ &= \tfrac{1}{2}D + \tfrac{1}{64}H + \tfrac{1}{5}(\tfrac{1}{4}D + \tfrac{1}{32}H + E_b) + \tfrac{1}{25}(\tfrac{1}{8}D + \tfrac{1}{16}H + E_w). \end{aligned}$$

The coefficients of D, H, E_w and E_b so obtained are set out in columns 5-8 of the upper part of Table 15. P_1, P_2 and the reciprocal F_1's were each raised as five plots in each block. Thus not only could an estimate of $E_1 = E_w$ be obtained from the pooled variances of parents and F_1's within plots; but an estimate of E_2, the non-heritable variance between plots, can also be found from the pooled variances between plot means, taken round the block means. In addition to E_b this will include an item of $\tfrac{1}{5}E_w$ because of sampling variation resulting from the variances within plots.

The direct estimates of E_1 and E_2, together with the variance of F_2, the two variances from F_3 and the three from F_3 are shown in Table 15, which also gives the number of degrees of freedom (df) on which each variance is based. (The details of the derivation of their number of degrees of freedom are given by M and J.) There are thus eight observed statistics from which we must estimate four parameters, D, H, E_w and E_b. This will leave four degrees of freedom for testing the goodness of fit of the model.

The procedure is essentially the same method of weighted least squares already described for the analysis of means (page 38). One difference must, however, be noted. The variances of means, whose reciprocals are used as weights in the analysis, are commonly observed empirically in the experiments. Replication is, however, seldom sufficient to permit the use of the same procedure where variances themselves are to be analysed, and in consequence the theoretical variance of the variance must be used to supply the reciprocals for use as weights. The variance of a variance V is $2V^2/N$, where N is the number of degrees

TABLE 15.

Analysis of Hayman's (1960) experiment on plant height in *Nicotiana rustica*

	Observed	df	First weight	Coefficients of D	H	E_w	E_b
V_{1F2}	69.29	80	0.008 331	0.500	0.250 000	1.00	0
V_{1F3}	43.12	36	0.009 681	0.550	0.087 500	0.20	1.0
V_{2F3}	36.66	160	0.059 526	0.250	0.125 000	1.00	0
V_{1F4}	67.84	36	0.003 911	0.555	0.024 375	0.04	0.2
V_{2F4}	41.29	153	0.044 872	0.275	0.043 750	0.20	1.0
V_{3F4}	26.47	770	0.549 481	0.125	0.062 500	1.00	0
From V_{P1}, V_{P2} and V_{F1}: E_1	12.95	160	0.477 035	0	0	1.00	0
From V_{P1}, V_{P2} and V_{F1}: E_2	14.06	32	0.080 937	0	0	0.20	1.0

	Expectation after iteration 1	2	5		Estimate after iteration 1	2	5
V_{1F2}	64.49	64.87	65.07	$\hat{D}$	79.98	99.01	97.51
V_{1F3}	61.80	68.37	68.16	$\hat{H}$	44.93	8.67	12.63
V_{2F3}	38.88	39.04	39.12	$\hat{E}_w$	13.27	13.20	13.16
V_{1F4}	48.26	57.79	57.11	$\hat{E}_b$	11.23	10.52	10.79
V_{2F4}	37.84	40.76	40.79	s_D	20.04	17.40	19.31
V_{3F4}	26.07	26.12	26.14	s_H	50.64	45.08	48.37
E_1	13.27	13.20	13.16	s_{Ew}	1.43	1.35	1.35
E_2	13.88	13.16	13.42	s_{Eb}	3.65	3.07	3.06
				$\chi^2_{[4]}$	5.87	3.67	3.68
				$\sqrt{(H/D)}$	0.75	0.30	0.36

of freedom from which V is estimated. These variances of variances should, however, be found using not the values observed for V_{1F2}, etc., but the values expected for them based on the estimates of D, H, E_w and E_b obtained by the weighted analysis. In other words finding the best estimates of the components of variation depends on using weights which themselves depend on the estimates of the components obtained using correct weights. We therefore proceed by the process of iteration, calculating the weights, first from the observed values of V_{1F2}, etc. These weights are used to obtain estimates of D, etc. which are in turn used to find expected values for V_{1F2} etc. New weights are computed from the expected values of the statistics and the process repeated until further repetition fails to improve the estimates and the test of goodness of fit. In the case of Hayman's experiment, two rounds of iteration are sufficient to achieve this result.

The values observed for the statistics are set out in the second column of Table 15, from which the first weights used in the first round of calculations, can be found as shown in column 3. Thus for V_{1F2}, its variance is

$$\frac{2V^2}{N} = \frac{2 \times 69.29^2}{80} = 120.028$$

and the first weight is $\dfrac{1}{120.028} = 0.008\,331.$

We then proceed to find the **J** and **S** matrices using these weights and the coefficients, of D, H, E_w, and E_b exactly as in the earlier example except that since there are now four parameters there will be four equations of estimation (not three as in the earlier example) with the consequence that **J** will be a 4×4 matrix and **S** a 4×1 matrix. Solution of the four equations of estimation, by finding $\mathbf{J}^{-1}\mathbf{S}$, gives the estimates of D, H, E_w and E_b shown in the second column of the lower right-hand portion of the table, their standard errors being obtained by taking the square roots of the four values in the leading diagonal of $\mathbf{J}^{-1}$. The values expected for V_{1F2} etc. are computed using these estimates of D, etc. and $\chi^2_{[4]}$ testing goodness of fit with the model is found in exactly the same way as in the earlier example. This χ^2 has four degrees of freedom since four parameters have been estimated from the eight observed statistics. The $\chi^2_{[4]}$ is not significant even in the first test and there is thus no indication that the model is inadequate.

Weights for the second iteration are found from the values of V_{1F2} etc., expected after the first iteration. In the case of V_{1F2}, its expected value is 64.49, giving as its variance $(2 \times 64.49^2)/80$ and for the second weight $80/(2 \times 64.49^2) = 0.009\,618$. Only the weights for V_{1F3} and V_{1F4}

change substantially, in the case of V_{1F3} from 0.009 681 (first weight) to 0.004 713 (second weight) and for V_{1F4} from 0.003 911 to 0.007 729. Nevertheless, when a new round of estimation is carried out exactly like the first calculation except that the new weights are used instead of the earlier ones, the estimates of D, and especially H are substantially changed, although E_w and E_b are not materially affected. New expectations can then be found for V_{1F2} etc. as shown in the lower left portion of the table and $\chi^2_{[4]}$ calculated to test the goodness of fit. This now turns out to be $\chi^2_{[4]} = 3.67$ with a probability of 0.30. Again there is clearly no indication of inadequacy of the model: indeed the fit is now better than after the first iteration. The new expectation for V_{1F2} can be used to find a third set of weights leading to a third round of calculations, and the process continued as long as one wishes. Hayman actually carried out five iterations, and the results of the fifth are shown in the table. It is clear that nothing was gained by continuing beyond the second round of calculations.

The standard errors of D, H, E_w and E_b are shown in the lower right portion of the table. s_H is large, so large indeed that there is no good evidence that H departs from 0, i.e. no good evidence of dominance. Nor should we be surprised at this when we see how low the coefficients of H are in the composition of the various statistics found from the experimental data: dominance clearly contributes relatively little to variation in the types of family raised in this experiment and we should therefore expect the estimate of H to be imprecise. It is for this reason too that the estimate of H changes so much more than those of D, E_w and E_b as we proceed from the first to the second iteration, and it will indeed be observed that despite the apparently large size of the change in the estimate of H it is not in fact large when compared with s_H. If a prime aim of the experiment had been to investigate the dominance properties of the genes, it would clearly have been desirable to include in it some types of families to whose variation dominance made greater contributions: indeed the inclusion of back-crosses would of itself have materially improved the estimate of dominance effects since H contributes as much as D to variation in such families.

A type of experiment especially well suited to the detection and measurement of dominance by the partitioning of variation is the so-called North Carolina Design III (M and J). It has the further advantage of leading to a simple analysis of variance, as does Kearsey and Jinks' (1968) triple test cross, which is an extension of N.C.D.III capable also of testing for interaction between non-allelic genes. Valuable as these are in particular respects, N.C.D.III and similar types of experiment are however of restricted use, as they suffer from two major limitations. In the first place only certain types of family can be utilized in them, and the number of variances obtainable from them is so restricted that little can be done towards testing the validity of the assumption that the genes contribute independently to the variation under investigation. Secondly, the analysis of variance, to which such designs lead, offers no means of combining several different generations into a single analysis, and so of multiplying the number of statistics available for use in estimating D, H and E in the way necessary not only for testing their adequacy as a represention of the variation but also for estimating the further components of variation that, as we shall see in later chapters, may be necessitated when the variation has a more complex structure than is provided by the simple additive-dominance model. The great merits of analysis by weighted least squares, illustrated by Hayman's experiment, are that it leads directly to a test of the adequacy of the model, that it is completely flexible in regard to the generations and types of family whose statistics can be brought into the analysis and that it is completely general in that it can be extended to cover structures and models of variation of any degree of complexity.

One final point remains to be noted about Hayman's experiment. He made no use of the covariances W_{1F23}, W_{1F34} and W_{2F34} that the F_2, F_3 and F_4 can yield in addition to their variances. Furthermore, his F_3 families were obtained by selfing F_2 plants other than those which he measured for the purpose of finding V_{1F2} and his F_4's were obtained by selfing F_3 plants other than those from whose measurements the F_3 variances were obtained. In this way he could ensure that, being based on unrelated plants, the variances from different generations were uncorrelated. Suppose, however, the same F_2 plants had been used for taking the measurements from which V_{1F2} was found and for raising the F_3s. A sampling correlation between V_{1F2} and V_{1F3} would have resulted. Also if W_{1F23} had been calculated from the same F_2 measurements and F_3 means, it too would have shown a sampling correlation with both the variances. In such a case the weights used in calculating the estimates of the components of variation can no longer be the simple reciprocals of the variances of V_{1F2} etc., but must take into account the sampling covariances of the statistics. A procedure is available for dealing with these more complicated applications of the method of weighted least squares (see M and J). No new basic principles are involved since the simpler analysis we have described is just a special case of the more general approach, but the necessary calculations become much heavier. Whereas the analysis of an experiment, like Hayman's, designed to avoid the complication of sampling correlation between the statistics, can be carried out without any great trouble on an electronic desk-calculator, the analysis of results where the statistics are subject to sampling correlations is virtually impracticable without access to an electronic computer.

4

Diallels

15. The principles of diallel analysis

Consider two true-breeding lines which differ in the alleles they bear at a locus, A-a, one thus being AA and the other aa. If they are mated in all possible combinations the four progenies so produced will of course consist of two which are like the two parents respectively and two which are the reciprocal F_1s. These four families can be arranged according to their parentage as in Table 16, which also shows the respective phenotypes

TABLE 16.

The four families obtained by mating two true-breeding lines differing in one gene, A-a

Male parent	Female parent AA d	Female parent aa $-d$	Mean 0
AA d	AA d	aA h	$\frac{1}{2}(d+h)$
aa $-d$	Aa h	aa $-d$	$\frac{1}{2}(h-d)$
Mean	$\frac{1}{2}(d+h)$	$\frac{1}{2}(h-d)$	$\frac{1}{2}h$
V_r	$\frac{1}{4}(d-h)^2$	$\frac{1}{4}(d+h)^2$	$\frac{1}{4}(d^2+h^2)$
W_r	$\frac{1}{2}d(d-h)$	$\frac{1}{2}d(d+h)$	$\frac{1}{2}d^2$

expressed as deviation from the mid-parent value, m. The table is symmetrical round its leading diagonal, each male array (row) having a common male parent, being like the female array (column) which has the same genotype as its common female parent. The table also gives the mean and variance (V_r) in respect of this gene for each array. It will be seen that the array variances, like the variances of back-crosses, will differ only if dominance is present. A further statistic can be calculated for each array. This is W_r, the covariance of the family means within the array with the phenotypes of their respective non-recurrent parents. Thus for the array whose common parent is AA, $W_r = \frac{1}{2}d_a \cdot d_a + \frac{1}{2}(-d_a)h_a = \frac{1}{2}d_a(d_a - h_a)$. Again W_r is the same for both arrays in the absence of dominance. The mean variance of the arrays is $\bar{V}_r = \frac{1}{2}[\frac{1}{4}(d_a - h_a)^2 + \frac{1}{4}(d_a + h_a)^2] = \frac{1}{4}(d_a^2 + h_a^2)$ and the mean covariance is similarly $\bar{W}_r = \frac{1}{2}d_a^2$. The variance of the array means can also be found as

$$V_{\bar{r}} = \tfrac{1}{2}[\tfrac{1}{2}(d_a + h_a)]^2 + \tfrac{1}{2}[\tfrac{1}{2}(h_a - d_a)]^2 - (\tfrac{1}{2}h_a)^2 = \tfrac{1}{4}d_a^2$$

and $V_{\bar{r}} + \bar{V}_r = \frac{1}{4}d_a^2 + \frac{1}{4}(d_a^2 + h_a^2) = \frac{1}{2}d_a^2 + \frac{1}{4}h_a^2$ which equals the contribution of such a gene difference to V_{1F2} (Table 12), as indeed it obviously should since an F_2 includes AA, Aa and aa individuals in the same proportions as the families of the corresponding genotypes in Table 16.

We can take the analysis further by considering the relation between W_r and V_r. Now the difference between the variances of the two arrays is $\Delta V_r = \frac{1}{4}[(d_a + h_a)^2 - (d_a - h_a)^2] = d_a h_a$ and that between the covariance is $\Delta W_r = \frac{1}{2}d_a[(d_a + h_a) - (d_a - h_a)] = d_a h_a$. Thus if we plot W_r against V_r as in Fig. 7, the line joining the two points must have a slope of $d_a h_a / d_a h_a$

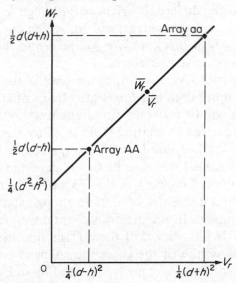

Fig. 7. The W_r/V_r graph, neglecting non-heritable variation, from a diallel set of matings involving one gene difference, A-a, where $h = \frac{1}{3}d$. The line passing through two points, from arrays AA and aa respectively, also passes through the point $\bar{W}_r$, $\bar{V}_r$ and has a slope of 1. It cuts the ordinate at $W_r = \frac{1}{4}(d^2 - h^2)$.

= 1 and it will pass through the point $\overline{W}_r$, $\overline{V}_r$, which as we have seen will be the point $\frac{1}{2}d_a^2$, $\frac{1}{4}(d_a^2 + h_a^2)$. So, if we project the line passing through the two points of the figure backwards it will cut the ordinate, where $V_r = 0$, at the value of W_r given by

$$\overline{W}_r - \overline{V}_r = \tfrac{1}{2}d_a^2 - \tfrac{1}{4}(d_a^2 + h_a^2) = \tfrac{1}{4}(d_a^2 - h_a^2).$$

The relative position of the two array points on the line will reflect the direction of dominance. If the A allele is dominant, that is h_a is positive, the point for array 1 (common parent AA) will occupy the lower position on the line. If, however, the a allele is dominant and h_a negative the point for array 2 (common parent aa) will occupy the lower position on the line. This graph therefore tells us a great deal about the genetical situation. In the absence of dominance, V_r is the same for both arrays and so is W_r. The two points on the graph will thus coincide except for random sampling variation in the estimates of V_r and W_r. If they do not coincide, the intercept on the ordinate of the line which joins them will provide a measure of dominance, and in particular where $h_a < d_a$ it will cut the ordinate above the origin, where $h_a = d_a$ it will pass through the origin and where $h_a > d_a$ it will pass below the origin. It should be noted, of course, that so far we have neglected non-heritable variation, which will contribute to the different variances (although in a suitably designed experiment not to the covariances) and for which due allowance must be made in any analysis of this kind. We will return to the nature of the necessary allowances at a later stage.

If the two true-breeding lines which are used as the parents of the families differ at more than one locus the effects of all the genes by which they differ will be reflected simultaneously in the phenotypes of the four families derived by mating them in all four possible combinations. In other words d_a and h_a must be replaced by $[d]$ and $[h]$. The information to be gained will thus be of the same kind as that obtainable from an analysis of means (Section 8) and being restricted to parental and F_1 families it will not even yield enough statistics to test the adequacy of the model. In the previous chapter we examined the limitations of $[d]$ and $[h]$ in respect of the information they provide about the dominance properties of the genes they depend on. We saw too how these limitations can be overcome by proceeding to F_2 and other segregating generations, which in addition to providing the additional means needed to test the adequacy of the model also yield second degree statistics enabling us to estimate and bring into the interpretation the quadratic quantities $D = S(d^2)$ and $H = S(h^2)$. We will now examine an alternative approach.

Table 16 is the simplest example of a diallel set of mating in which a number, n, of true-breeding lines are mated together in all possible combinations to give n^2 families. Since it involved only two lines ($n = 2$) it could clearly give us information about only one genetical difference, or, if more than one such difference was involved, only about the differences as a unitary aggregate. If more lines are used, clearly a correspondingly greater number of differences, or aggregate differences, can be investigated. As the next simplest case let us consider a diallel among four lines representing all the possible combinations of two gene differences, A-a and B-b. The genotypes of the 16 families so obtained are shown in Table 17 as are the phenotypes expected on the assumption that A-a

TABLE 17.

Diallel set of matings involving four true-breeding lines, being all the combinations of two genes, A-a and B-b

Male parent	Female parent				Mean
	AABB d_a+d_b	AAbb d_a-d_b	aaBB $-d_a+d_b$	aabb $-d_a-d_b$	0
AABB d_a+d_b	AABB d_a+d_b	AABb d_a+h_b	AaBB h_a+d_b	AaBb h_a+h_b	$\frac{1}{2}[(d_a+h_a) + (d_b+h_b)]$
AAbb d_a-d_b	AABb d_a+h_b	AAbb d_a-d_b	AaBb h_a+h_b	Aabb h_a-d_b	$\frac{1}{2}[(d_a+h_a) + (h_b-d_b)]$
aaBB $-d_a+d_b$	AaBB h_a+d_b	AaBb h_a+h_b	aaBB $-d_a+d_b$	aaBb $-d_a+h_b$	$\frac{1}{2}[(h_a-d_a) + (d_b+h_b)]$
aabb $-d_a-d_b$	AaBb h_a+h_b	Aabb h_a-d_b	aaBb $-d_a+h_b$	aabb $-d_a-d_b$	$\frac{1}{2}[(h_a-d_a) + (h_b-d_b)]$
Mean	$\frac{1}{2}[(d_a+h_a) + (d_b+h_b)]$	$\frac{1}{2}[(d_a+h_a) + (h_b-d_b)]$	$\frac{1}{2}[(h_a-d_a) + (d_b+h_b)]$	$\frac{1}{2}[(h_a-d_a) + (h_b-d_b)]$	$\frac{1}{2}(h_a+h_b)$
V_r	$\frac{1}{4}[(d_a-h_a)^2 + (d_b-h_b)^2]$	$\frac{1}{4}[(d_a-h_a)^2 + (d_b+h_b)^2]$	$\frac{1}{4}[(d_a+h_a)^2 + (d_b-h_b)^2]$	$\frac{1}{4}[(d_a+h_a)^2 + (d_b+h_b)^2]$	$\frac{1}{4}(d_a^2+h_a^2+d_b^2+h_b^2) = \frac{1}{4}(D+H)$
W_r	$\frac{1}{2}[d_a(d_a-h_a) + d_b(d_b-h_b)]$	$\frac{1}{2}[d_a(d_a-h_a) + d_b(d_b+h_b)]$	$\frac{1}{2}[d_a(d_a+h_a) + d_b(d_b-h_b)]$	$\frac{1}{2}[d_a(d_a+h_a) + d_b(d_b+h_b)]$	$\frac{1}{2}(d_a^2+d_b^2) = \frac{1}{2}D$

and B-b contribute independently. At the foot of the table are the four V_r's one for each array, and similarly the four W_r's. It will be observed that, as in the earlier example, $\Delta W_r = \Delta V_r$ when we move from one array to another. Thus moving from array AAbb to AABB gives $\Delta W_r = \Delta V_r = d_b h_b$, and from aabb to AABB gives $\Delta W_r = \Delta V_r = d_a h_a + d_b h_b$. So, if we plot W_r against V_r the four points, one from each array, will lie on a straight line of slope 1. Furthermore it must pass through the point $\overline{W}_r, \overline{V}_r$ which is $\frac{1}{2}(d_a^2 + d_b^2), \frac{1}{4}(d_a^2 + h_a^2 + d_b^2 + h_b^2)$ and may be rewritten as $\frac{1}{2}D, \frac{1}{4}(D + H)$. The line will thus cut the ordinate at $\overline{W}_r - \overline{V}_r = \frac{1}{2}D - \frac{1}{4}(D + H) = \frac{1}{4}(D - H)$. So we can learn something of the average dominance relations of the two genes and indeed, bearing in mind that the variance among the four parent means is $V_P = \frac{1}{4}[(d_a + d_b)^2 + (d_a - d_b)^2 + (-d_a + d_b)^2 + (-d_a - d_b)^2] = d_a^2 + d_b^2 = D$, we can obtain an estimate of the average dominance as $\sqrt{[(V_P - 4I)/V_P]} = \sqrt{(H/D)}$, where I is the intercept of the regression line with the ordinate.

We should note, too, that now two genes, A-a and B-b, are involved the relation of W_r to V_r provides a test of the additive-dominance model of gene action. The phenotypes set out in Table 17 are those expected when the two gene pairs make independent contributions to the expression of the character. If their contributions are not independent, that is if the genes interact in producing their effects, we cannot expect the relation of W_r and V_r, to hold good as we have derived them, and in particular we can no longer expect the regression of W_r on V_r to be rectilinear with a slope of 1.

16. An example of a simple diallel

An example will illustrate how the diallel analysis and the test of the additive-dominance model can be carried out in practice. The data are taken from a larger experiment carried out using the eight substitution lines between the Wellington and Samarkand inbred lines of *Drosophila melanogaster* to which we referred in Chapter 1. The character followed was again sternopleural chaeta number. The results of mating four of the substitution lines WWW, WWS, WSW and WSS in all combinations are shown in Table 18. Since the X chromosome was the same in all four parent lines it can be ignored and the lines will thus be designated by their contributions in respect of chromosomes II and III. It will be recalled from Chapter 1 that all the substitution lines were homozygous for their respective chromosomes. The set of sixteen matings was duplicated, a complete set being raised on each of two occasions, and the duplicates are recorded separately in the table, each entry of which is the mean number of chaeta from five female and five male progeny.

The observations of Table 18 may be subjected to an analysis of variance. The $16 \times 2 = 32$ observations had 31df of which 15 will be for differences among the 16 matings, 1 for the overall difference between

TABLE 18.

Sternopleural chaeta number in a diallel set of matings among four true-breeding lines, being all the combinations of the Wellington (W) and Samarkand (S) chromosomes II and III in *Drosophila melanogaster*, made on two occasions. The two entries in each cell of the table are one from each of the two occasions

Male parent	Female parent				Mean
	WW	WS	SW	SS	
WW	17.45	17.25	18.20	17.65	17.9000
	17.65	18.35	18.45	18.20	
WS	18.05	18.80	18.10	18.85	18.5940
	18.55	18.80	18.45	19.15	
SW	17.40	18.40	19.05	18.50	18.6313
	18.40	19.00	19.40	18.90	
SS	17.95	18.95	18.65	19.10	18.6500
	17.15	18.85	18.95	19.60	
Mean	17.8250	18.5500	18.6563	18.7438	18.4438

the sets of matings reared on the two occasions, and 15 for the interaction of matings × occasions, i.e. for the differences between the duplicate observations after allowance has been made for the overall difference between occasions. The 15 df for differences between matings may be partitioned into 3 items, namely 3 df for differences among the 4 genotypes of female parents, 3 for differences among the 4 genotypes of male parents, and 9 for the interaction of female and male parental genotypes. The main items for differences among female and male parents both reflect differences among the same set of four genotypes and so, in the absence of complications such as maternal effects, should yield estimates of the same component of variation, which will of course be the additive variation (D). The item for interaction of female and male parents will test for departures from simple additivity of the gene effects, including dominance as well as non-additivity of non-allelic genes in producing their effects. The analysis of variance is set out in Table 19. The matings × occasions item provides an estimate of the error variation. The mean square for occasions is significant, so confirming that, as might be expected, the experimental conditions were not pre-

TABLE 19.

Analysis of variance of the diallel data in Table 18

Item	df	MS	VR	P
Female parents	3	1.411 46	15.02	<0.001
Male parents	3	1.055 63	11.23	<0.001
Interaction	9	0.283 06	3.01	0.05 − 0.01
Occasions	1	0.945 31	10.06	0.01 − 0.001
Matings × Occasions (Error)	15	0.093 98		
Reciprocals	6	0.128 65	1.37	>0.20

cisely the same at the times when the progenies were raised from the duplicate sets of matings. The mean squares for the differences among the four genotypes are significant for both female and male parents, showing that there is additive genetic variation among these genotypes. The item for interaction of the differences among the female and male parents, although not so large, is also significant, so showing that the differences among the sixteen progenies are not wholly accountable in terms of additive variation: there must also be present non-additive variation to which both dominance and interaction of non-allelic genes could contribute.

The mean squares for female parents and male parents do not differ significantly from one another, as would be expected if the two sexes are contributing equally to the genotypes of the progeny. There is thus no indication of any maternal effect, or of indeed any other departure from simple autosomal inheritance, and the close comparison of means for the corresponding arrays from female parents and male parents shown in the margins of Table 18 confirms this. A further and more stringent test is, however, possible. The four matings along the leading diagonal of the diallel table (Table 18) are repeats of the homozygous parental lines, the female and male parents being of the same genotype in each case. The other twelve matings are between parents of different genotypes and fall into six pairs of reciprocal crosses. Provided the parents contribute equally to the progeny these reciprocals should be alike within the limits of sampling variation. The mean square for differences between reciprocals can thus be compared with error variation to provide a test of equilinearity in the genetical determination of the character. The mean square is readily found. Thus the duplicate progenies from WS × WW gives values of 17.25 and 18.35 while those from WW × WS give 18.05 and 18.55. The difference between the reciprocals is therefore 17.25 + 18.35 − 18.05 − 18.55 = −1.0 and the contribution of this comparison to the sum of squares (SS) is $(-1.0)^2/4$, the divisor 4 reflecting the use of 4 observations in deriving the difference. There are 6 such differences, obtained from the 6 pairs of reciprocal crosses, as set out in Table 20, and summing their contributions yields a SS of 0.771 875. This

TABLE 20.

Differences between the offspring of reciprocal crosses in the data of Table 18

	WS	SW	SS
WW	−1.00	0.85	0.75
WS		−0.85	0.20
SW			−0.20
	Total	−0.25	

SS stems from 6 comparisons and so takes 6 df, thus yielding a MS of $\frac{1}{6}(0.771\,875) = 0.128\,65$ as shown below the main analysis in Table 19. This MS does not depart significantly from the estimate of error variation and there is hence no evidence of any departure from simple autosomal inheritance. The 6 df included in this test are part of the 15 df for differences among matings and represent a partition of these 15 different from the partition used in the main analysis, and testing a different feature of the genetical situation. More comprehensive analyses of variance of the diallel tables are available, notably one by Hayman. These test a wider range of features, but are more complex to carry out. They will therefore not be described here, but a full account of Hayman's analysis of variance is given by M and J.

We have now established that there is not only additive variation, but non-additive also, between the four genotypes, and that there is no evidence of reciprocal differences. We can proceed to analyse the non-additive variation further, and in particular to test whether dominance is adequate to account for it or whether interaction of non-allelic genes must also be invoked, by examining the relations between W_r and V_r. Since there is no evidence of differences between the progenies of reciprocal crosses, we can combine these to give single values for each cross between different lines, and we can of course also pool the values from duplicate progenies. This gives us the reduced or half-diallel table shown in Table 21. The entries along the diagonal of this table are for

TABLE 21.
Half-diallel table from the data of Table 18

	WW	WS	SW	SS	Mean	W_r	V_r
WW	*17.5500*	18.0500	18.1125	17.7375	17.8625	0.1427	0.0703
WS		*18.8000*	18.4875	18.9500	18.5719	0.2748	0.1582
SW			*19.2250*	18.7500	18.6438	0.3232	0.2186
SS				*19.3500*	18.6969	0.5271	0.4713
					Mean	0.3169	0.2296

the progenies of mating within the four parental genotypes and thus are repeats of these four parental lines. Each is the mean of two duplicate progenies. Thus for WW × WW we have $\frac{1}{2}(17.45 + 17.65) = 17.55$. The off-diagonal entries on the other hand are the means of four progenies, namely the pair of reciprocals each of which is represented by duplicate progenies. Thus the entry for WW × WS is $\frac{1}{4}(17.25 + 18.35 + 18.05 + 18.55) = 18.05$. In proceding to find W_r and V_r we note that, after pooling our reciprocals, it does not matter whether we work on female or male arrays: they will give identical results. The WS array for example, consists of WW × WS (18.0500), WS × WS (18.8000), WS × SW (18.4875) and WS × SS (18.9500) as shown by the linking lines in Table 21. Its V_r is thus $\frac{1}{3}[(18.0500^2 + 18.8000^2 + 18.4875^2 + 18.9500^2) - \frac{1}{4}(18.0500 + 18.8000 + 18.4875 + 18.9500)^2] = 0.1582$ the final divisor being 3 because there are 3 df among the 4 progenies. These values of V_r are entered in the right-hand column of the table.

The calculation of W_r requires a further word of explanation. We could have used values for the four parental lines obtained from progenies of these lines obtained independently of the diallel itself. This is, however, unnecessary as the four parental lines appear along the leading diagonal of the diallel table and we can in fact utilize these four entries in the table to provide values of the mean chaeta numbers of the four parental genotypes. (This introduces a complication in assessing the values of the components of variation, as we shall see later (p. 80), but one which does not affect our immediate analysis and so may be ignored for the moment.) So again taking the WS array as an example, we find its W_r as

$$\tfrac{1}{3}[(18.0500 \times 17.5500) + (18.8000 \times 18.8000) + (18.4875 \times 19.2250) + (18.9500 \times 19.3500)] - \tfrac{1}{4}(18.0500 + 18.8000 + 18.4875 + 18.9500)(17.5500 + 18.8000 + 19.2250 + 19.3500)] = 0.2748.$$

The values of W_r for the four arrays are given next to those for the corresponding V_r in Table 21.

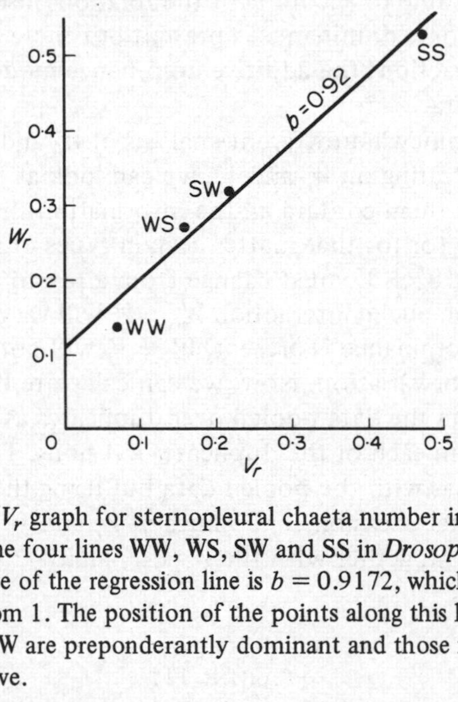

Fig. 8. The W_r/V_r graph for sternopleural chaeta number in the defined diallel among the four lines WW, WS, SW and SS in *Drosophila melanogaster.* The slope of the regression line is $b = 0.9172$, which does not differ significantly from 1. The position of the points along this line shows that the genes from W are preponderantly dominant and those from S preponderantly recessive.

If we now plot W_r against V_r (Fig. 8) we expect to find a straight line of slope 1 if the non-additive variation is wholly ascribable to dominance. The regression of W_r on V_r can be calculated in the customary way. There are 3 df among the four points, one from each array, and we find $SS(W_r) = 0.076\,297$, $SCP(W_r, V_r) = 0.081\,464$ and $SS(V_r) = 0.089\,010$. Then $b = \dfrac{0.081\,464}{0.089\,010} = 0.9152$ and it accounts for $\dfrac{0.081\,464^2}{0.089\,010} = 0.074\,558$ of the $SS(W_r)$ leaving 0.001 739 as the remainder SS for deviations from the regression line. Since the assignment of 1 df to the regression line leaves 2 df for the remainder SS, the error variation against which the regression SS must be tested is $\frac{1}{2}(0.001\ 739) = 0.000\ 870$, and $t_{[2]}$ testing the significance of the slope of the regression line is $\sqrt{(0.074\ 56/0.000\ 870} = 9.26$ which even with no more than 2 df for the estimate of error, has a probability of only 0.01. Clearly there is a significant regression of W_r on V_r. Furthermore the standard error of b will be found as $\sqrt{(\text{error variance}/\text{SS}(V_r))} = \sqrt{(0.000\ 870/0.089\ 010)} = 0.0988$ and it is clear that the value of does not depart significantly from 1. Thus, so far as this analysis goes there is good evidence of dominance, but no evidence that dominance is not wholly able to account for the relation observed between W_r and V_r. In other words dominance is present but there is no indication of non-allelic interaction: the additive-dominance model is sufficient to account for the data.

A further and somewhat different analysis of W_r and V_r is possible. Instead of concentrating on W_r and V_r, we can look at $W_r + V_r$ and $W_r - V_r$, which between them contain all the information that W_r and V_r carry. Now if dominance (or for that matter certain types of non-allelic interaction) are present $W_r + V_r$ must change from array to array. At the same time, if there is non-allelic interaction $W_r - V_r$ will vary between arrays, although if only dominance is present, $W_r - V_r$ will not vary more than expected from error variation. Now, we can calculate W_r and V_r for each array not only from the data pooled over duplicates as we did above, but also separately from each of the duplicate occasions. The calculation is, of course, exactly as with the pooled data but using the separate data from each individual occasion. The values of W_r and V_r so obtained are shown in Table 22, together with the $W_r + V_r$ and $W_r - V_r$ derived from

TABLE 22.

Values of W_r and V_r from the two occasions

Array	Occasion 1				Occasion 2			
	W_r	V_r	W_r+V_r	W_r-V_r	W_r	V_r	W_r+V_r	W_r-V_r
WW	0.1242	0.0275	0.1517	0:0967	0.1283	0.2004	0.3288	−0.0721
WS	0.3750	0.3317	0.7067	0.0433	0.1772	0.0518	0.2290	0.1254
SW	0.3467	0.2781	0.6427	0.0686	0.2914	0.1677	0.4590	0.1237
SS	0.4063	0.3268	0.7331	0.0794	0.6696	0.6538	1.3233	0.0158

them. There are thus eight values for each of $W_r + V_r$ and $W_r - V_r$, one from each of the four arrays in each of the two halves of the experiment. We can now carry out an analysis of variance on $W_r + V_r$ and another similarly on $W_r - V_r$. In each case there will be 7 df among the 8 observed values, of which 3 can be ascribed to differences between the arrays and the remaining 4 to the differences between the duplicate values obtained for each of the 4 arrays. These 4 df could be further partitioned into 1 df for the overall difference between occasions and 3 df for variation of the 4 array differences round this overall value; but this is unnecessary in the present case since the overall difference between occasions is not significant when compared with the residual variation for the 3 df. We thus have a simple analysis into two parts, one of which for 4 df is a measure of the variation within arrays between occasions and provides the estimate of error against which the mean square between arrays can be tested for significance.

The two analyses of variance, for $W_r + V_r$ and $W_r - V_r$ respectively, so obtained are set out in Table 23. The MS between arrays for $W_r - V_r$ is

TABLE 23.

Analyses of variance of W_r+V_r and W_r-V_r

	Item	df	MS	
W_r+V_r	Between arrays	3	0.2200	VR = 2.77
	Within arrays	4	0.0794	P = 0.20−0.05
W_r-V_r	Between arrays	3	0.0029	Not
	Within arrays	4	0.0053	significant

not significant when tested against that within arrays and indeed is smaller than it. There is thus no evidence of any non-allelic interaction; no evidence, that is, of any inadequacy of the additive-dominance model. Turning to the analysis of variance $W_r + V_r$, it will be seen that the MS between arrays is greater than that within them, but not significantly so. On this evidence alone, therefore, we could not be confident that even dominance was present. We should recall, however, the evidence from the initial analysis of variance (Table 19) of non-additive effects, which must be accounted for in some way. Since there is no evidence of interaction between non-allelic genes, we must conclude that although not formally significant by itself the higher value for the MS between arrays for $W_r + V_r$, does in fact reflect dominance, and that while the assumption of additive genetic variation alone is not adequate, the additive-dominance model does provide an adequate basis for interpreting the results.

Returning to the overall estimates of W_r and V_r obtained when the data from the two halves of the experiment are pooled (Table 21), their mean values are $\overline{W}_r = 0.3169$ and $\overline{V}_r = 0.2296$. To these two statistics we may add the variance of the parent lines (V_P) found from the leading diagonal of the diallel table which, as has already been noted, comprises the four parental genotypes. We thus find from Table 21 $V_P = \frac{1}{3}[17.550^2 + 18.800^2 + 19.225^2 + 19.350^2 - \frac{1}{4}(17.550 + 18.800 + 19.225 + 19.350)^2] = 0.675\,573$. However, before we can use these estimates for deriving the values of the genetical components of variation D and H, they must be corrected for the non-heritable items that they contain. The original analysis of variance of the experiment (Table 19) yielded a value of 0.093 98 for the error variance based on the differences between the duplicate observations made on each of the sixteen matings in the table. This error variation reflects, of course, the non-heritable differences to which the observations are subject and hence provides the basis for finding the non-heritable components of the three statistics in which we are now interested. We note that each value along the leading diagonal of Table 21 is the mean of a pair of duplicate observations. These will thus be subject to half the error variation of the single observations and we can estimate the non-heritable component of V_P, which is the variance of the values in this leading diagonal, as $\frac{1}{2} \times 0.093\,98 = 0.046\,99$. Thus the heritable component of $V_P = D = 0.675\,57 - 0.046\,99 = 0.628\,58$. The off-diagonal entries in Table 21 are, however, the means of four observations each, and so will be subject to only $\frac{1}{4}$ the error variation of single observations. V_r for each array is based on three such off diagonal entries together with one diagonal entry. In other words $\frac{3}{4}$ of the observations on which V_r is based are each subject to $\frac{1}{4}$ of the error variance, and $\frac{1}{4}$ of the observations are subject to $\frac{1}{2}$ the error variance. Thus the non-heritable component of each V_r, can be estimated as $(\frac{3}{4}\cdot\frac{1}{4} + \frac{1}{4}\cdot\frac{1}{2})$ $0.093\,98 = 0.029\,37$ and the heritable component of $V_r = \frac{1}{4}(D + H) = 0.229\,59 - 0.029\,37 = 0.200\,22$.

Turning to $\overline{W}_r$, we note that it would contain no non-heritable item if it had been calculated using values of the parental lines from observations made independently of the diallel matings. In fact, however, we are taking the parental values from the leading diagonal of the diallel table itself. So, every W_r will include, as one of the four cross-products from which it is derived, the square of the appropriate parental value. Thus, for example, as we have already seen, W_r for the WS array is based on

$$(18.0500 \times 17.5500) + (18.8000)^2 + (18.4875 \times 19.2250) + (18.9500 + 19.3500).$$

This squared value will bring in an item for non-heritable variation. It is a value from the leading diagonal of Table 21 and so is the mean of two observations and it provides one of the four cross-products that contribute to each W_r. Hence the non-heritable component of W_r will be $(\frac{1}{4} \cdot \frac{1}{2})$ 0.093 98 = 0.011 75 and the genetic component of $\overline{W}_r = \frac{1}{2}D$ thus be-becomes 0.316 93 − 0.011 75 = 0.305 18. Before proceeding we might observe that while the regression of W_r on V_r used in analysing their relationship should strictly be the regression of the genetic portion of W_r on the genetic portion of V_r, the regression of the W_r on V_r uncorrected for their non-heritable components (as used in Fig. 8) will give exactly the same value for b since we subtract a common non-heritable item from all four W_r and also a common one from all four V_r. The slope of the regression line is thus not affected, even although its position as defined by the point $\overline{W}_r$, $\overline{V}_r$, through which it must pass, and hence its intercept with the ordinate, is valid only after the non-heritable components have been deducted.

Returning to our main theme, there is another statistic which we have not used so far but which can be calculated from the diallel table, namely the variance of array means, $V_{\bar{r}}$ whose heritable component is $\frac{1}{4}D$. These means are shown in Table 21 from which we find $V_{\bar{r}}$ = 0.152 78. This variance, too, will contain a non-heritable component. Each array mean is derived from an array as shown in Table 21, and thus corresponds to the joint mean of the corresponding female and male arrays of Table 18: in fact the mean of the WW array is the mean of all the observations in the first column and first row of Table 18, the observations in the top left corner each having been used twice. The array mean is thus the mean of twelve observations used once each and two used twice, thus being the equivalent of $(12 \times 1) + (2 \times 2) = 16$ observations. But when an observation is multiplied by two, the amount it contributes to a variance is multiplied by four. So the non-heritable component of the variance of array sums will be $(12 \times 1) + (2 \times 4) = 20$ times the error variance and the non-heritable variance of array means $V_{\bar{r}}$, will be correspondingly

$$\frac{20}{16^2}\,(0.093\,98) = 0.007\,34.$$

So after deducting the non-heritable components we have

$$V_P = D = 0.628\,58, \quad \bar{V}_r = \tfrac{1}{4}(D+H) = 0.200\,22$$
$$\bar{W}_r = \tfrac{1}{2}D = 0.305\,18, \quad V_{\bar{r}} = \tfrac{1}{4}D = 0.145\,44$$

We can thus find estimates of D and H as

$$D = \tfrac{4}{7}(V_P + \bar{W}_r + V_{\bar{r}}) = \tfrac{4}{7}\,(1.079\,20) = 0.616\,69$$
$$H = 4\bar{V}_r - D = 0.800\,88 - 0.616\,69 = 0.184\,19.$$

Then as an estimate of the average level of dominance we can take

$$\sqrt{(H/D)} = \sqrt{\frac{0.184\,19}{0.616\,69}} = \pm 0.546\,51.$$

These results are collected together in Table 24.

TABLE 24.

Components of variation in the diallel of Table 18

	Total	Non-genetic	Genetic	Expectation
V_P	0.6756	0.0470	0.6286	D
$\bar{W}_r$	0.3169	0.0117	0.3052	$\frac{1}{2}D$
$V_{\bar{r}}$	0.1528	0.0073	0.1455	$\frac{1}{4}D$
$\bar{V}_r$	0.2296	0.0294	0.2002	$\frac{1}{4}(D+H)$

$$D = \tfrac{4}{7}(V_P + \bar{W}_r + V_{\bar{r}}) = 0.6167$$
$$H = 4\bar{V}_r - D = 0.1842$$
$$\sqrt{\frac{H}{D}} = 0.5465$$

We might note that this procedure for estimating D and H is not fully efficient as we have given V_P, $\bar{W}_r$ and $V_{\bar{r}}$ equal weight in finding D. A more complex procedure can be used to provide least squares estimates which at $D = 0.622\,88$, $H = 0.177\,92$ and $\sqrt{(H/D)} = 0.534\,45$ are virtually identical with those yielded by the simpler procedure.

While we now have an estimate of the dominance ratio we do not as yet have any indication as to its direction. But as we have earlier noted, the order of the points on the W_r, V_r graph itself gives an indication of the relative number of dominant to recessive genes present in the common parent of each array: the common parent with the most dominant genes has the smallest values of W_r and V_r and that with the most recessive genes the largest values of W_r and V_r. Now it can be seen from Table 21 that the order of the arrays from the smallest to the largest values of W_r and V_r is WW, WS, SW and SS. Since WW has a smaller value than SW and WS than SS, the W chromosome II must show dominance over its S homologue. Similarly WW gives smaller values for W_r and V_r than does WS, and SW than SS. Thus the W homologue of chromosome III also shows dominance over the S. Since, therefore, the W homologues are also associated with a lower score (Table 21) and the S homologues with a higher score, the direction of dominance is clearly prepondantly for lower score, the dominance deviations being negative.

This diallel is defined in the sense that the genotypes of the parents, and hence of the progenies, are known for every mating. It is therefore possible to approach its analysis in a different way. The progenies of the sixteen matings fall into the nine genotypes expected for all the possible combinates of two 'genes' each with two 'alleles'. The different genotypes are not expected to be produced by the same number of matings; the homozygotes WWWW, WWSS, SSWW, and SSSS are each represented by single matings, although of course duplicate progenies are available for each of them; the four single heterozygotes WWWS, WSWW, SSWS, and WSSS each came from two matings (reciprocals) and so are represented by four progenies; and the double heterozygote is produced by four matings (those along the off-diagonal in Table 18) and so is represented by eight progenies. The mean chaeta number of the nine genotypes, obtained by averaging over the appropriate observations in Table 18 are set out in Table 25, together with (in brackets) the number of observations from which each is derived. The means in the margins of the table are the means of all flies of the particular genotype in question. Thus, for example, the mean of all flies homozygous for the W chromosome II is given at the bottom of the first column having been found as $\frac{1}{8}[(17.55 \times 2) + (18.05 \times 4) + (18.80 \times 2)] = 18.1125$. The expected departures of these marginal means from the mid-parent value of the whole experiment are also shown in terms of d_2, d_3, h_2 and h_3, where the subscripts 2 and 3 refer to chromosomes II and III respectively. As will be readily seen, we can estimate these four parameters from the marginal means, the chromosome II parameters from the lower margin of the table and the chromosome III from the right-hand margin. Considering the chromosome II parameters

$$d_2 = \tfrac{1}{2}[(d_2 + \tfrac{1}{2}h_3) - (-d_2 + \tfrac{1}{2}h_3)] = \tfrac{1}{2}(19.0188 - 18.1125] = 0.4531$$

$$h_2 = \tfrac{1}{2}[2(h_2 + \tfrac{1}{2}h_3) - (d_2 + \tfrac{1}{2}h_3) - (-d_2 + \tfrac{1}{2}h_3)] = \tfrac{1}{2}[2 \times 18.3219 - 19.0188 - 18.1125] = -0.2438.$$

TABLE 25.

Direct estimation of genetic parameters

Chromosome III	Chromosome II W/W	W/S	S/S	Mean	Expectation
W/W	17.5500 (2)	18.1125 (4)	19.2250 (2)	18.2500	$m+\frac{1}{2}h_2-d_3$
W/S	18.0500 (4)	18.1125 (8)	18.7500 (4)	18.2563	$m+\frac{1}{2}h_2+h_3$
S/S	18.8000 (2)	18.9500 (4)	19.3500 (2)	19.0125	$m+\frac{1}{2}h_2+d_3$
	18.1125	18.3219	19.0188	18.4438	
	$m-d_2+\frac{1}{2}h_3$	$m+h_2+\frac{1}{2}h_3$	$m+d_2+\frac{1}{2}h_3$		$m+\frac{1}{2}h_2+\frac{1}{2}h_3$

$m = 18.7532$

	Chromosome II	III	Mean	Diallel analysis
d	0.4531	0.3813	0.4172	0.5553
h	−0.2438	−0.3750	−0.3094	−0.3035
$\frac{h}{d}$	−0.5381	−0.9835	−0.7410	−0.5466

These values and those yielded similarly for d_3 and h_3 by the right-hand marginal means are shown in the lower part of Table 25. It will be observed that for both chromosomes the S homologue mediated a higher chaeta number than W, and also that h is negative in both cases, indicative that the W homologue is showing dominance over S for both chromosomes II and III. Now $D = d_2^2 + d^2_3 = 0.4531^2 + 0.3813^2 = 0.3507$ which compares with the estimate $D = 0.6167$ obtained from the diallel analysis, and similarly $H = h_2^2 + h_3^2 = 0.2001$ as compared with the estimate $H = 0.1842$ from the diallel analysis. The agreement between the two estimates of H is close, but that between the two estimates of D less so. We should remember, however, that D and H are quadratic quantities and hence will tend to magnify apparent discrepancies. In order to make a comparison in linear quantities, let us note that the direct estimates of d_2 and d_3 do not differ significantly and hence assume that they are equal. Similarly h_2 and h_3 do not differ significantly and we assume that they also are equal. We then replace d_2 and d_3 each by their mean $\bar{d}$, and h_2 and h_3 similarly by $\bar{h}$. The values for $\bar{d}$ and $\bar{h}$ obtained from the direct analysis are shown in the column headed Mean in the lower part of Table 25. Turning to the estimates from the diallel analysis, $D = 2\bar{d}^2$ and $H = 2\bar{h}^2$. Then $\bar{d} = \sqrt{(\frac{1}{2} \cdot 0.6167)} = 0.5553$ and $\bar{h} = \sqrt{\frac{1}{2} \cdot 0.1842)} = -0.3035$, these findings being entered in the column of the table headed Diallel Analysis. That h from the diallel analysis must in fact have a negative sign is shown, as already noted, by the order of the points on the W_r, V_r graph.

The agreement in respect of h is now strikingly good and that in respect of d reasonably close. In fact, although it is not easy to test the significance of the difference between the two estimates of d, it is unlikely to be significant. If we now estimate the average level of dominance by taking h/d we obtain -0.7416 from the entries in the mean column and -0.5466 from the diallel column. The two analyses agree in showing dominance to be incomplete, lying somewhere between half and three-quarters, and in the direction of low chaeta number. Evidently the diallel analysis has produced estimates which are compatible with those of the direct analysis, over and above it showing that while dominance is present there is no evidence for interaction of non-allelic genes.

17. Undefined diallels

Just as a 4 × 4 diallel can be used to investigate two genetic differences, in the way we have seen, an 8 × 8 could be designed using as parents all the possible combinations of three genetic differences and used to examine the behaviour of these genetic differences and to test whether they showed non-allelic interaction. We could go on to a 16 × 16 to look at four genetic differences in the same way, and so on. Where, however, the genotypes of the parents, and hence of the progenies are defined and known, as in the case we have described, the approach through direct analysis is always open and will in general lead to more informative results since the d's and h's are then estimated individually and not pooled in D and H. The value of applying the diallel analysis to the experiment discussed in the last section, was in fact that it allowed us to compare its results with those of direct analysis and see that it did effectively extract the same information.

With the vast majority of diallels, direct analysis is not possible because it is rare for the parental genotypes to be defined as they were in the *Drosophila* experiment. Where the differences among the parental genotypes are undefined, diallel analysis must be used and two further complications must immediately be taken into account. In the first place we cannot know that the two alleles (assuming that there are only two) of any gene are equally common among the parents, other than in exceptional cases like the diallel referred to by Jinks *et al.* (1969) in which the 20 parental lines were descended by selfing from 20 individuals in an F_2 of *Nicotiana rustica* and hence might be expected to have equal frequencies for the alleles at any locus, within the limits of sampling variation.

Secondly, we cannot be sure either that the pairs of alleles at different loci are distributed at random with respect to each other in the way that can be ensured in a defined diallel. Clearly we must take the possibility of such association of the genes into account in carrying out the analysis and interpreting its results.

Let us look into the consequences of these complications, starting with that of unequal gene frequencies. Consider the case where a proportion u_a of the parent lines are true-breeding for allele A and proportion $v_a (= 1 - u_a)$ are true-breeding for allele a. The mating AA and AA will then occur in u_a^2 of cases and of aa with aa in v_a^2 of cases, the remaining $2u_a v_a$ of matings being AA × aa. The frequencies of the types of matings, together with the genotypes and phenotypes in respect of this gene difference are shown in Table 26. The array means, variances and covariances are also shown in the table. Just one point needs noting about their derivation. The mating AA × AA, for example, constitutes $u_a \times u_a = u_a^2$ of all matings in the table, but it constitutes u_a of the matings in the arrays stemming from AA parents. Thus the mean of the AA array is $u_a d_a + v_a h_a$, not $u^2 d_a + v^2 h_a$. Bearing the same point in mind, the variance of that same array is found as

$$V_r = u_a d_a^2 + v_a h_a^2 - (u_a d_a + v_a h_a)^2 = u_a v_a (d_a - h_a)^2$$

and the covariance is

$$W_r = u_a d_a \cdot d_a - v_a d_a \cdot h_a - (u_a - v_a) d_a (u_a d_a + v_a h_a) = 2u_a v_a d_a (d_a - h_a).$$

The mean, V_r and W_r of the aa array are found similarly.

We can then see from the table that the changes in V_r and W_r between the arrays are respectively

$$\Delta V_r = 4u_a v_a d_a h_a \quad \text{and} \quad \Delta W_r = 4u_a v_a d_a h_a.$$

Thus inequality of the frequencies of the alleles A and a makes no dif-

TABLE 26.

Diallel set of matings where u of the parents are homozygous for allele A, and $v(=1-u)$ are homozygous for allele a

		Female parent		
	Genotype	AA	aa	Mean
	Frequency	u	v	
	Expression	d	$-d$	$(u-v)d$
Male parent	AA	AA	Aa	
	u	u^2	uv	
	d	d	h	$ud+vh$
	aa	Aa	aa	
	v	uv	v^2	
	$-d$	h	$-d$	$uh-vd$
	Mean	$ud+vh$	$uh-vd$	$(u-v)d+2uvh$
	V_r	$uv(d-h)^2$	$uv(d+h)^2$	$uv[d+(v-u)h]^2+4u^2v^2h^2 = \frac{1}{4}(D_R+H_R)$
	W_r	$2uvd(d-h)$	$2uvd(d+h)$	$2uvd[d+(v-u)h] = \frac{1}{2}D_W$

$$V_P = 4uvd^2 = D_P$$

$$\Delta W_r = 4uvdh \quad \Delta V_r = 4uvdh$$

ference to two important properties of W_r and V_r. First the arrays will have the same V_r and W_r in the absence of dominance, i.e. when $h_a = 0$ the arrays will all give the same point, within the limits of sampling variation, on the W_r, V_r graph. Secondly, where $h_a \neq 0$ the slope of the line joining the points from the two arrays on the W_r/V_r graph will have a slope of 1. It will be observed that if $u_a = v_a$, i.e. if the frequencies of A and a are equal, all these expressions reduce to those found for the simple case discussed at the beginning of the Chapter, as indeed they clearly should.

In extending our consideration to two genetic differences, we note that where the frequencies of A and a are u_a and v_a respectively among the true-breeding parents, and the frequencies of B and b are similarly u_b and v_b, the alleles at the two loci will be distributed independently of each other if the frequency of AB, Ab, aB and ab parents are $u_a u_b$, $u_a v_b$, $v_a u_b$ and $v_a v_b$ respectively. Given that this is the case, and assuming that the effects of non-allelic genes are simply additive, that is that there is no non-allelic interaction, it is not difficult to derive the expression for the array means, variances and covariances shown in Table 27. These expressions reduce of course to those in Table 17 when $u_a = v_a = u_b = v_b = \frac{1}{2}$. It will be seen that for any pair of arrays $\Delta V_r = \Delta W_r$. Thus for example the differences between arrays aabb and AABB are $\Delta V_r = \Delta W_r = 4(u_a v_a d_a h_a + u_b v_b d_b h_b)$ while those between arrays aaBB and AAbb are $\Delta V_r = \Delta W_r = 4(u_a v_a d_a h_a - u_b v_b d_b h_b)$. Thus in plotting their W_r against V_r the four arrays will again give four points lying in a straight line of slope 1, and also again the array with the two dominant alleles will have the lowest values of W_r and V_r, and so will give the lowest point on the graph while the array with the two recessive alleles will give the highest point with the other two arrays giving intermediate points. Thus the test of adequacy of the additive-dominance model developed for the defined diallel in the previous section will apply to undefined diallels. We should note, however, that an undefined diallel will reveal failure of the model not only when the genes show non-allelic interaction, i.e. are not independent in their action but also when the genes show non-random association among the parents, i.e. are non-independent in their distribution. Finally, it is not difficult to see that these relations between W_r and V_r, and with them the test of goodness of fit of the additive-dominance model, still hold good for three, four or indeed any number of gene differences. They are in fact general properties of diallel sets of matings.

So far nothing has been said about the genetical components of variation D and H, and indeed when we turn to these we find complexities which were not present in the case of the defined diallel. Turning back to the case of the single gene difference in an undefined diallel (Table 26) we find that the contribution this pair of alleles makes to V_P the variance of the parents is no longer d_a^2, but takes the more general form $4u_a v_a d_a^2$, which of course becomes d_a^2 when the alleles are equally frequent among the parents of the diallel, i.e. $u_a = v_a = \frac{1}{2}$. With two genes independent in their actions and their distribution $V_P = 4u_a v_a d_a^2 + 4u_b v_b d_b^2$ and with any number of genes $V_P = \mathrm{S}(4uvd^2)$. We may thus write $V_P = D_P$ where $D_P = \mathrm{S}(4uvd^2)$.

When we turn to array variances, however, while the contributions of A-a to $\overline{V}_r$ may still be written as the sum of two quadratic quantities, one of which depends solely on h^2, the other no longer depends solely on d^2. The contribution to $\overline{V}_r$ is in fact $u_a v_a [d_a + (v_a - u_a)h_a]^2 + 4u_a^2 v_a^2 h_a^2$ and generalizing to any number of genes independent in their actions and their distribution the genetical component of $\overline{V}_r = \mathrm{S}\{uv[d + (v-u)h]^2 + 4u^2 v^2 h^2\}$. This can be cast in the form $\overline{V}_r = \frac{1}{4}(D_R + H_R)$

TABLE 27.

Array frequencies, means, variances and covariances for two gene differences, A-a with frequencies u_a and v_a, and B-b with frequencies u_b and v_b

	Array				Overall
	AABB	AAbb	aaBB	aabb	
Frequency	$u_a u_b$	$u_a v_b$	$v_a u_b$	$v_a v_b$	1
Mean	$u_a d_a + v_a h_a$ $+ u_b d_b + v_b h_b$	$u_a d_a + v_a h_a$ $+ u_b h_b - v_b d_b$	$u_a h_a - v_a d_a$ $+ u_b d_b + v_b h_b$	$u_a h_a - v_a d_a$ $+ u_b h_b - v_b d_b$	$S[(u-v)d + 2uvh]$
V_r	$u_a v_a (d_a - h_a)^2$ $+ u_b v_b (d_b - h_b)^2$	$u_a v_a (d_a - h_a)^2$ $+ u_b v_b (d_b + h_b)^2$	$u_a v_a (d_a + h_a)^2$ $+ u_b v_b (d_b - h_b)^2$	$u_a v_a (d_a + h_a)^2$ $+ u_b v_b (d_b + h_b)^2$	$S[uv(d + \overline{v-u}h)^2 + 4u^2 v^2 h^2]$ $= \frac{1}{4}(D_R + H_R)$
W_r	$2u_a v_a d_a (d_a - h_a)$ $+ 2u_b v_b d_b (d_b - h_b)$	$2u_a v_a d_a (d_a - h_a)$ $+ 2u_b v_b d_b (d_b + h_b)$	$2u_a v_a d_a (d_a + h_a)$ $+ 2u_b v_b d_b (d_b - h_b)$	$2u_a v_a d_a (d_a + h_a)$ $+ 2u_b v_b d_b (d_b + h_b)$	$S[2uvd(d + \overline{v-u}h)]$ $= \frac{1}{2}D_W$

$$V_P = 4(u_a v_a d_a^2 + u_b v_b d_b^2) = D_P$$

when we use the definitions $D_R = S\{4uv[d + (v - u)h]^2\}$ and $H_R = S(16u^2v^2h^2)$ which again reduce to the standard forms $D = S(d^2)$ and $H = S(h^2)$ when $u = v = \frac{1}{2}$. We shall meet these components D_R and H_R again. The covariances are different again for $\bar{W}_r = \frac{1}{2}D_W$ where $D_W = S\{4uvd\,[d + (v - u)h]\}$. It will be observed that the individual contribution $4uvd[d + (v - u)h]$ to D_W is the geometric mean of the contributions $4uvd^2$ to D_P and $4uv\,[d + (v - u)h]$ to D_R. This is not surprising when we recall that $\bar{W}_r$ is the average covariance of offspring, whose average variance is $\bar{V}_r$, with their non-recurrent parents, whose variance is V_P.

Thus the simple assessment of the components of variation that was possible with the defined diallel is no longer so with the undefined. The very differences in the definitions of D and H as they appear in V_P, $\bar{V}_r$ and $\bar{W}_r$ can, however, be turned to profit by the use of a more complex analysis of the relations between V_P, $\bar{V}_r$ and $\bar{W}_r$ which can not only yield a measurement of average dominance of the form $\sqrt{[S(h^2)/S(d^2)]}$, but also a measure of the average value of uv and hence of the disimilarity in the frequencies of alleles, and even under certain circumstances of the relative frequencies of dominant and recessive alleles. This analysis, which also brings in the variance of array means and covariance, $V_{\bar{r}}$ and $W_{\bar{r}}$, would however, take us beyond the scope of the present discussion. It is set out fully in M and J's discussion of diallels.

18. An example of an undefined diallel

The number of parent lines in a defined diallel is rigidly fixed by the number of combinations of the genes involved: thus with two genes the number of parents is four, with three genes it is eight, and so on. In undefined diallels on the other hand there is no such restriction on the number of parent lines and indeed any number can be used. The values of u and v will reflect the frequencies of the two alleles in the actual set of parents chosen, and although the frequencies of the different combinations of genes among the parents cannot generally agree precisely with the frequencies u_au_b, u_av_b, v_au_b, v_av_b and so on expected from independent distributions of the genes, provided the departures fall within sampling variation the assumption of independent distribution will be sufficiently well realized for the diallel analysis to proceed without disturbance arising from non-independence of the distributions.

As an example of the analysis of an undefined diallel we will take the 9 × 9 diallel from *Nicotiana rustica* quoted by M and J. The parents were nine inbred lines, the character was date of opening of the first flower in days after 1st July (the choice of date for the origin is of no consequence as a change of it merely alters the mean of the experiment), and the experiment was laid out as 2 blocks, each of which consisted of 81 plots to which the 81 progenies of the 9 × 9 matings were assigned at random. Each plot comprised 5 plants and the datum from each plot is the mean flowering time of the 5 plants it contained. The flowering time of each of the 81 progenies, averaged over the 2 blocks, is shown in Table 28. There was no overall difference between the 2 blocks in respect of flowering time, and all 81 df for differences between the duplicate progenies in the 2 blocks may therefore be used in the estimate of error, their mean square being 3.858.

The analysis proceeds in a way exactly analogous to that of the defined diallel in Section 16. The analysis of variance corresponding to that in Table 19 may be carried out from the data in Table 28, bearing in mind that since the observations in this table are the means of duplicate plots, the SS found from them must be multiplied by 2 to put them on the single plot basis. There will be 80 df among the 81 progenies, 8 for differences among the 9 female parents, 8 for differences among the male parents and 8 × 8 = 64 for interaction. The analysis of variance is set out in Table 29, and all these items, for female parents, male parents and interactions, are highly significant when tested against the duplicate error variance of 3.858.

There are 36 pairs of reciprocal crosses and we can therefore find a SS corresponding to 36 df for differences between reciprocals, in just the same way that we did in the earlier example. This turns out to give a MS of 290.092/36 = 8.058, which has a probability of between 0.01 and 0.001 when tested against the duplicate error. It must therefore be regarded as significant. We cannot, however, regard it as clearly demonstrating an extra-nuclear element in the determination of flowering time: true it could reflect such a determinant, but it could arise in other ways too. For example, if the inbreeding of the parent lines had not been completely effective and some residual variation remained in them, and if precisely the same parent plants had not been used in making the reciprocal crosses, differences such as those observed could have arisen. Equally if the seed for each family was sown in a single seed pan, members of a family, including those plants grown in separate blocks as well as those in the same block, could resemble one another more than

TABLE 28.

Flowering-time in a 9 X 9 diallel set of matings in *Nicotiana rustica*.
All entries are means of duplicate observations from two blocks.
The flowering-times of the parental lines are in italics

Male parent	Female parent									Mean
	1	2	3	4	5	6	7	8	9	
1	*38.90*	26.70	39.80	34.80	25.10	29.80	35.70	33.80	25.30	32.2111
2	23.90	*27.05*	25.00	23.10	21.50	26.20	23.40	20.60	20.20	23.4389
3	34.40	26.60	*48.80*	29.55	25.00	31.50	36.10	24.40	26.00	31.3722
4	36.10	23.50	31.20	*34.10*	23.40	29.35	27.20	22.30	25.00	28.0167
5	26.50	23.20	26.00	25.50	*26.60*	27.50	27.20	20.20	24.20	25.2111
6	28.40	24.10	30.30	31.90	24.15	*27.00*	27.70	22.40	24.80	26.7500
7	36.90	24.70	41.80	33.90	30.10	29.80	*37.00*	24.40	29.10	31.9667
8	26.80	19.30	27.80	22.10	19.20	18.80	22.70	*15.30*	21.80	21.5333
9	25.30	23.30	24.90	24.00	22.50	21.30	27.40	19.00	*25.40*	23.6778
Mean	30.8000	24.2722	32.8444	28.7722	24.1722	26.8056	29.3778	22.4889	24.6444	27.1309

TABLE 29.

Analysis of variance of the 9 × 9 diallel in *Nicotiana rustica*

Item	df	MS	Duplicate – Error – Reciprocal			
			VR	P	VR	P
Female parents	8	221.876	56.8	v.s.	27.5	v.s.
Male parents	8	289.541	74.1	v.s.	35.9	v.s.
Interaction	64	20.923	5.4	<0.001	2.6	0.001
Duplicate error	81	3.858				
Reciprocals	36	8.058				

v.s. = very small

they resembled the plants from the reciprocal crosses started off in a different seed pan. This would produce the result observed and later experiments in fact pointed to it as the most likely cause. Whatever the explanation, however, it is clear that the duplicate error variance is not a reliable yardstick to use in assessing the significance of the items in the analysis of variance. When tested against the reciprocal mean square, 8.058, the probabilities of the variances for female and male parents are still very small and even that for interaction still has a probability as low as 0.001. Thus even when tested against this new and higher estimate of error, all the items are still significant.

It should be noted that this test of significance is not strictly valid, since the 36 df for reciprocal differences are not orthogonal to the 3 items, female parents, male parents and interaction, contained in the 80 df that these 3 items jointly comprise. Since, however, the reciprocals mean square is lower than any of the other 3, deduction of the 36 df from the 80 could only serve to raise the mean square attaching to the residual df's and so raise the VR and hence the significance. Although our test is not strictly valid, it is a conservative test and we can therefore accept the significance that it reveals for all 3 items in the main analysis of variance. The Hayman analysis of variance of these data described by M and J overcomes this difficulty and confirms the conclusions from this simple analysis.

The significant interaction item in the analysis of variance shows us that there is non-additive heritable variation, and we must now continue the analysis to discover whether this non-additive element can be accounted for by dominance or whether non-independence of the effects of non-allelic genes is also involved. Proceeding just as we did in the earlier example, the values of $W_r + V_r$ and $W_r - V_r$, taken from M and J, are listed for each of the nine arrays from each of the two blocks in the upper part of Table 30, with their analyses of variance in the lower part

TABLE 30.

$W_r + V_r$ and $W_r - V_r$ from the two blocks, 1 and 2

Array	$W_r + V_r$		$W_r - V_r$	
	1	2	1	2
1	81.939	61.763	13.728	3.419
2	25.105	14.731	11.650	6.141
3	159.529	112.761	9.055	2.106
4	80.289	41.610	7.453	9.125
5	29.814	19.128	14.277	4.919
6	51.130	36.211	21.844	15.149
7	91.263	72.303	20.668	16.212
8	55.178	55.213	19.008	16.410
9	29.152	15.093	15.535	4.691

Analyses of variance		$W_r + V_r$			$W_r - V_r$		
	df	MS	VR	P	MS	VR	P
Between arrays	8	2736.0	9.67	<0.001	50.34	1.96	0.20–0.10
Within arrays	9	282.9			25.95		

of the table. As in the *Drosophila* example, we have not taken out the single degree of freedom for the block difference because our analysis of the original data again shows no evidence of such a difference. It is clear that $W_r + V_r$ varies significantly from array to array whereas $W_r - V_r$ does not. There is therefore clear evidence of dominance but no evidence of non-independence in effect of non-allelic genes. This means that not only is there no evidence of interaction between non-allelic genes in producing their effects, but also that there is no evidence of the genes being associated in a non-random way in their distributions between the parents.

We can move on to the regression of W_r on V_r. The arrays pooled over blocks and reciprocals are set out in Table 31. The values of W_r and V_r are also shown for each array. The SS for W_r is 2680.75 and for V_r is 2892.76, while the SCP for W_r and V_r is 2690.30. The linear regression of W_r on V_r is thus $b = 2690.30/2892.76 = 0.9300$ which does not differ signifi-

TABLE 31.

Half-diallel table pooled over the two blocks, with array means, V_r and W_r

	Arrays									Mean	V_r	W_r
	1	2	3	4	5	6	7	8	9			
1	38.90	25.30	37.10	35.45	25.80	29.10	36.30	30.30	25.30	31.506	30.1665	38.9886
2		27.05	25.80	23.30	22.35	25.15	24.05	19.95	21.75	23.856	5.0059	13.4255
3			48.80	30.38	25.50	30.90	38.95	26.10	25.45	32.108	64.8375	70.3335
4				34.10	24.45	30.63	30.55	22.20	24.50	28.394	23.8661	33.6539
5					26.60	25.83	28.65	19.70	23.35	24.692	6.8419	16.0434
6						27.00	28.75	20.60	23.05	26.778	12.0930	30.3851
7							37.00	23.55	28.25	30.672	31.0601	49.6899
8								15.30	20.40	22.011	18.3805	35.8031
9									25.40	24.161	5.3974	15.8965

cantly from 1, although of course it departs very significantly from 0. Again there is clear evidence of dominance, but no evidence of non-independence in the effects of non-allelic genes. Evidently the additive-dominance model is adequate to account for the behaviour of this diallel.

The graph showing the regression of W_r on V_r is plotted in Fig. 9. The lowest point is from array 2 whose parent line must therefore carry the largest number of dominant alleles, while the highest is from array 3 which must carry the smallest number of dominant alleles. The other arrays give intermediate points whose order shows the order in numbers of dominant alleles carried by the parent. We can compare the value of $W_r + V_r$ for each array with the mean flowering time of the common parent of that array to see whether the distribution of dominant alleles is correlated with the phenotypes of the common parent. The parental flowering times ($\bar{P}$) are plotted against $W_r + V_r$ in Fig. 10, from which it is clear that in general the later flowering lines give the larger values of $W_r + V_r$ and so must be carrying fewer dominant genes. There is in fact a significant correlation of $r = 0.779$ between flowering time and $W_r + V_r$. Evidently the genes which give early flowering tend to be dominant. The anomalous position of line 8, which while being the earliest flowering of all the parents has an intermediate value of $W_r + V_r$, shows however that not all the genes for early flowering can be dominant and suggests that there is an ambidirectional element in the dominance relation of the flowering time genes, as would be expected if this character had been under stabilizing selection (Mather, 1973).

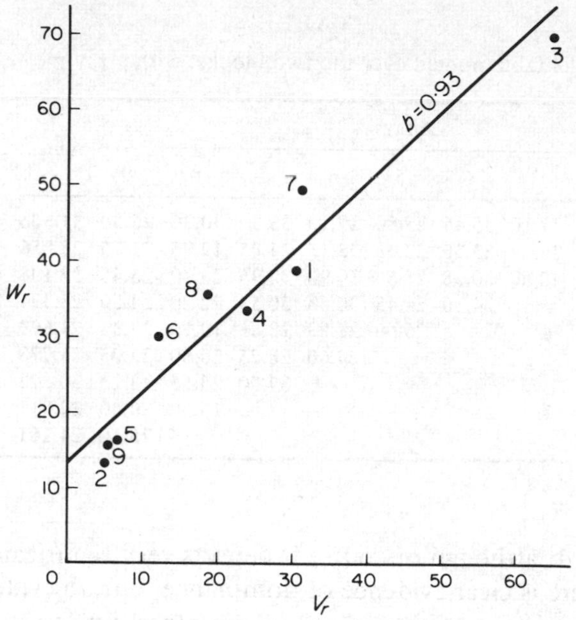

Fig. 9. The W_r/V_r graph for flowering time in the undefined diallel among nine lines of *Nicotiana rustica.* The parental line giving rise to the array represented by each point is indicated by the number against it.

Finally we turn to the components of variation. The values found for V_P, $\overline{W}_r$, $\overline{V}_r$ and $V_{\bar{r}}$ are listed in Table 32. Before they can be used for estimating the components of variation they must be corrected for their non-genetic components. The corrections are derived using the reciprocal mean square as the estimate of error, $V_E = 8.058$. The coefficients to be applied to V_E to find the corrections are obtained in the same way as in the earlier example, bearing in mind that there are now 9 parent lines and 9 items in each array, not 4 as in the earlier example. These coefficients are shown in the table, together with the actual values of the corrections and the resulting estimate of the heritable components of V_P, etc. V_P, $2\overline{W}_r$ and $4V_{\bar{r}}$ yield estimates of D_P, D_W and D_R respectively, while $4(\overline{V}_r - V_{\bar{r}})$ gives an estimate of H_R. Now $D_P = \mathrm{S}(4uvd^2)$, $D_W = \mathrm{S}\{4uvd\,[d + (v-u)h]\}$ and $D_R = \mathrm{S}\,\{4uv\,[d + (v-u)h]^2\}$. Thus each term in D_W, is the geometric mean of the corresponding terms in D_P and D_R. If the ratio of $[d + (v-u)h]$ to d is constant over all the genes D_W itself will then be the geometric mean of D_P and D_R, but if this ratio varies D_W must be less than $\sqrt{(D_P \cdot D_R)}$. In fact, as we see from Table 32 $\sqrt{(D_P \cdot D_R)} = \sqrt{(90.162 \times 53.947)} = 69.742$ while D_W is 66.709. The

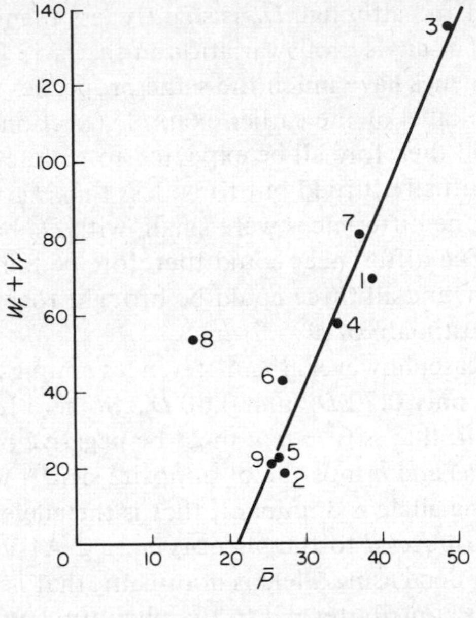

Fig. 10. $W_r + V_r$ from each array of the *Nicotiana rustica* diallel plotted against $\bar{P}$ the mean flowering time (expressed in days after 1st July) of the parental line giving rise to that array. Note that all the points lie as a straight regression line except that from parental line 8. With the exception of line 8, the earlier the flowering of the parent, the smaller the corresponding $W_r + V_r$, showing that in general the alleles for earlier flowering are dominant to those for later flowering. The position of point 8, however, indicates that this dominance relation no longer holds when the parent's flowering time is earlier than mid-July.

TABLE 32.

Components of variation for flowering-time in *Nicotiana rustica*

	Total	Non-genetic	Genetic	Expectation
V_P	94.1907	$\frac{1}{2}V_E = 4.0291$	90.1616	D_P
$\overline{W}_r$	33.8022	$\frac{1}{18}V_E = 0.4477$	33.3545	$\frac{1}{2}D_W$
$V_{\bar{r}}$	13.7355	$\frac{5}{162}V_E = 0.2487$	13.4868	$\frac{1}{4}D_R$
$\overline{V}_r$	21.9610	$\frac{5}{18}V_E = 2.2384$	19.7226	$\frac{1}{4}(D_R + H_R)$

$$
\begin{aligned}
D_P &= 4\,S[uvd^2] &&= V_P &&= 90.1616\\
D_W &= 4\,S[uvd(d+\overline{v-u}h)] &&= 2\overline{W}_r &&= 66.7090\\
D_R &= 4\,S[uv(d+\overline{v-u}h)^2] &&= 4V_{\bar{r}} &&= 53.9472\\
H_R &= 16\,S[u^2v^2h^2] &&= 4(\overline{V}_r - V_{\bar{r}}) &&= 24.9432
\end{aligned}
$$

agreement is good and although D_W is slightly less than $\sqrt{(D_P \cdot D_R)}$ there is little indication of any serious variation in the ratio $[d + (v - u)h]/d$. Evidently all the genes have much the same properties in this respect.

In the defined diallel of the earlier example (Section 16) all $u = v$, and D_P, D_W and D_R will therefore all be expected to yield estimates of $D = \mathrm{S}(d^2)$. While D_R actually turned out to be less than D_P with D_W intermediate in value, the differences were small, with D_R having a value over 0.92 that of D_P. The differences could therefore be fairly attributed to sampling variation, and all three could be brought together to give a common, overall estimate of $D = \mathrm{S}(d^2)$.

In the present case, however, the differences among D_P, D_W and D_R are marked: D_R is only 0.72 D_W and 0.60 D_P. So here $[d + (v - u)h]$ must be less than d, that is $(v - u)h$ must be negative or to put it in other words $(v - u)$ and h must be of opposite sign. h will be positive when the increasing allele is dominant, that is the allele which when homozygous contributes d to the phenotype (e.g. A), and h will be negative when the decreasing allele is dominant, that is the allele which when homozygous contributes $-d$ to the phenotype (e.g. a). Now u is the frequency of the increasing allele and v that of the decreasing allele, so that $v - u$ will be positive when the decreasing allele is the more common. So for $(v - u)h$ to be negative, the increasing allele must be more common (i.e. $v - u$ negative) when it is dominant (i.e. h positive); and equally the decreasing allele must be the more common ($v - u$ positive) when it is dominant (i.e. h negative). The conditions for $D_R < D_W < D_P$ are thus not only that dominance is present (otherwise all $h = 0$) and allele frequencies unequal (otherwise $v - u = 0$) but further that the dominant alleles are preponderantly more common than the recessives. Since in the present case the decreasing alleles, leading to earlier flowering times, must in the main be dominant, they must in general be more common than their increasing counterparts, or in other words since h is preponderantly negative, v must in general be greater than u.

It is possible to take the analysis still further and arrive at estimates of the average dominance (h/d) and of the average value of uv (and hence of u and v); but to do so would take us beyond the scope of our present discussion. The methods of doing so are set out by M and J.

5

Genic interaction and linkage

19. Non-allelic interaction

In the analyses of the foregoing chapters we have assumed that, save in one respect, the different genes were independent of one another in the contribution that they made to the various statistics, means, variances and covariances, under discussion. To put it in other words, we have assumed that the gene effects were simply additive, the exceptional respect being that we have accommodated dominance by incorporating the parameter h in the models we have constructed and tested.

Our analyses have included tests of the validity of this assumption - that, dominance apart, the genes were independent of each other in their contributions to the means, variances and covariances - in the form of, for example, scaling tests in the analysis of means or tests of the constancy of $W_r - V_r$ in diallel analysis. When subjected to these tests, the additive-dominance model by no means always proves to be adequate for the interpretation of the data: we must then conclude that the assumption of independence is invalid. Nor can we assume that the choice of a more appropriate scale on which to represent our measurements would always overcome the problem, for as we have seen earlier, in many cases a scalar transformation will clearly not serve to remove the difficulty. We need, therefore, means of explicitly accommodating the consequences of non-independence in the analysis.

Now the genes may show non-independence in two ways. First, they may be influenced by one another in their expression, i.e. they may interact in producing their effects. Secondly, they may be correlated with one another in their distribution among the individuals whose phenotypes are under investigation. We will consider interaction first, and to see how we proceed let us return for a moment to dominance.

In the absence of dominance individuals heterozygous for the gene Aa, would display a phenotype midway between those of the homozygotes AA and aa. The effect of substituting allele A for a would be independent of whether the allelic gene also present was A or a: the effects of the alleles would be simply additive and there would be no need to incorporate h into the model. The incorporation of h is at once a recognition that alleles need not be independent of each other in exerting their effects, and the provision of a parameter by which their interaction can be accommodated and measured. Dominance is thus the interaction of allelic genes and h is the parameter by which this allelic interaction is measured. We require corresponding means of representing and measuring the interaction of non-allelic genes, or non-allelic interaction as it is often called.

Consider the simplest case of two gene pairs, A-a and B-b. These can give rise to nine different genotypes each with its own phenotypic characteristics as shown in Table 33. The differences among these phenotypes

TABLE 33.

Phenotypes from the nine genotypes comprising all combinations of A-a and B-b

	AA	Aa	aa
BB	$d_a + d_b$ $+i_{ab}$	$h_a + d_b$ $+j_{ba}$	$-d_a + d_b$ $-i_{ab}$
Bb	$d_a + h_b$ $+j_{ab}$	$h_a + h_b$ $+l_{ab}$	$-d_a + h_b$ $-j_{ab}$
bb	$d_a - d_b$ $-i_{ab}$	$h_a - d_b$ $-j_{ba}$	$-d_a - d_b$ $+i_{ab}$

can therefore be completely described by eight parameters, which correspond of course to the 8 df among the nine observations. Four of these parameters we have already defined, namely d_a, d_b, h_a and h_b. The remaining four parameters can then be conveniently defined as representing respectively the interactions of d_a and d_b, d_a and h_b, h_a and d_b and h_a and h_b. Now d_a measures the difference in phenotype between AA and aa, and similarly d_b that between BB and bb. If d_a and d_b are independent, d_a will be the same whether the difference AA-aa is measured in BB or bb individuals. Thus with independence $AABB - aaBB = AAbb - aabb$ or $AABB - aaBB - AAbb + aabb = 0$, where $AABB$ is the phenotype of

AABB etc. Similarly in respect of d_b, $AABB - AAbb = aaBB - aabb$ or $AABB - aaBB - AAbb + aabb = 0$, as before. We can therefore accommodate prospective interaction of d_a and d_b by including a further parameter i_{ab} such that the phenotype of AABB is $d_a + d_b + i_{ab}$, that of AAbb is $d_a - d_b - i_{ab}$, that of aaBB is $-d_a + d_b - i_{ab}$ and that of aabb is $-d_a - d_b + i_{ab}$. Then the difference of AA and aa taken over both BB and bb genotypes is $(AABB - aaBB) + (AAbb - aabb) = 4d_a$ since the d_b's and i_{ab}'s cancel out. Similarly the overall difference of BB and bb is $(AABB - AAbb) + (aaBB - aabb) = 4d_b$, and the interaction of these differences is $(AABB - aaBB - AAbb + aabb) = 4i_{ab}$. The relation of these four completely homozygous classes has been described completely by the introduction of the new parameter i_{ab} representing the interaction of d_a and d_b. When there is no such interaction $i_{ab} = 0$ since $(AABB - aaBB - AAbb - aabb) = 4i_{ab} = 0$.

Turning to the relation of d_a and h_b, since d_a represents the difference between AA and aa, absence of interaction implies that h_b will be the same whether measured in individuals that are AA or individuals that are aa. In the presence of interaction between d_a and h_b, these two measurements will not be the same, and we can accommodate the interaction by including a new parameter j_{ab} such that it is added in the specification of AABb which is basically $d_a + h_b$, but subtracted in the specification of aaBb, which is basically $-d_a + h_b$. In the absence of interaction $j_{ab} = 0$, and its value provides a measure of any interaction that may be present between d_a and h_b. A corresponding parameter j_{ba} can be similarly incorporated into the specifications of AaBB and Aabb to represent and provide a measure of the interaction between h_a and d_b. The last of the four interactions, between h_a and h_b, is covered by a fourth parameter l_{ab} which is incorporated into the specification of AaBb, where h_a and h_b appear together.

The formulations of the phenotypes stemming from the nine genotypes are set out in terms of the eight parameters, 2 d's, 2 h's, i, 2 j's and l, in Table 33. The interaction terms are easy to derive: wherever the formulation includes d_a and d_b it also includes i; wherever it includes a d and an h it also includes the appropriate j; and wherever it includes h_a and h_b it includes l. In all cases the coefficient of the interaction term is the product of the coefficients of the two main items: thus $d_a + d_b$ is accompanied by i_{ab} whose coefficient is 1×1, while $d_a - d_b$ takes $-i_{ab}$ the coefficient being 1×-1, and so on. The system is readily extendable to the development of parameters covering trigenic and even more complicated interactions but we shall not now concern ourselves with these. One further point requires clarification. d and h were defined in Chapter 3 as deviations from the mid-parent m, that is mean of the two true-breeding parents from whose cross the families were derived. This definition of m can be seen now to be no longer adequate, for if we start with a cross between AABB and aabb, the mid-parent is

$$\tfrac{1}{2}[(m + d_a + d_b + i_{ab}) + (m - d_a - d_b + i_{ab})] = m + i_{ab}$$

whereas the alternative cross, AAbb × aaBB, gives a mid-parent of

$$\tfrac{1}{2}[(m + d_a - d_b - i_{ab}) + (m - d_a + d_b - i_{ab})] = m - i_{ab}$$

even although in the absence of linkage it gives just the same distribution of genotypes in F_2 and other derived generations as did AABB × aabb. In neither cross do the deviations cancel out and they leave residua which have opposite signs. The mid-parent must in fact be redefined as the mean of all the possible true-breeding combinations obtainable from the two gene pairs - in this case the mean of AABB, AAbb, aaBB and aabb which gives

$$\tfrac{1}{4}[(m + d_a + d_b + i_{ab}) + (m + d_a - d_b - i_{ab}) + (m - d_a + d_b - i_{ab}) + (m - d_a - d_b + i_{ab})] = m.$$

Before we proceed to discuss the use in analysis of these four interaction parameters, we should observe that since, together with the d's and h's, they afford a complete account of any differences that may be observed among the phenotypes of the nine genotypes, it follows that any system we may care to consider of interrelations among the nine phenotypes can be defined in terms of these parameters. Thus all the classical types of digenic interaction elucidated by Bateson and others in the early days of genetics can now be defined in biometrical terms. To take but two of the six classical interactions illustrated by Darlington and Mather (1949, Fig. 38), complementary gene action, first elucidated by Bateson and Punnett, gives a characteristic 9:7 ratio in F_2 because as Bateson and Punnett showed by breeding tests the genotypes AABB, AaBB, AABb and AaBb all had one phenotype, while AAbb, Aabb, aaBB, aaBb and aabb all had another. Allowing for the frequencies with which the genotypes appear in F_2, the first group will include $\frac{1}{16}(1 + 2 + 2 + 4) = \frac{9}{16}$ while the second will include $\frac{1}{16}(1 + 2 + 1 + 2 + 1) = \frac{7}{16}$ of the F_2 individuals. Writing these relations in our biometric notation, the likeness in phenotype of AABB, AABb, AaBB and AaBb requires that $d_a + d_b + i = d_a + h_b + j_a = h_a + d_b + j_b = h_a + h_b + l$, and the likeness of aaBB,

aaBb, AAbb, Aabb, and aabb requires that $-d_a + d_b - i = -d_a + h_b - j_a = d_a - d_b - i = h_a - d_b - j_b = -d_a - d_b + i$ where for the sake of convenience we write i_{ab} as i, j_{ab} as j_a, j_{ba} as j_b and l_{ab} as l. It is not difficult to show that these equations are satisfied if, and only if,

$$d_a = d_b = h_a = h_b = i = j_a = j_b = l.$$

Now in our usage, the designation of the commoner phenotypes as being produced by AABB, AaBB, etc. implies that this commoner phenotype is the one with the greater expression of the character. Clearly there could then be a counterpart situation where the phenotype with the lesser expression would constitute $\frac{9}{16}$ of the F_2, and that with the greater expression only $\frac{7}{16}$. This would arise where the phenotype of aabb, Aabb, aaBb and AaBb were alike on the one hand and those of aaBB, AaBB, AAbb, AABb and AABB were alike on the other. The equations then became

$$-d_a - d_b + i = h_a - d_b - j_b = -d_a + h_b - j_a = h_a + h_b + l$$

and

$$-d_a + d_b - i = h_a + d_b + j_b = d_a - d_b - i = d_a + h_b + j_a = d_a + d_b + i.$$

These equations are satisfied when $d_a = d_b = -h_a = -h_b = -i = j_a = j_b = -l$. Thus the general conditions for classical complementary action are that all eight parameters are equal in size, with two j's positive like the d's and i and l having the same sign as the two h's, which themselves are of the same sign.

The second classical interaction we will consider is that of so-called duplicate genes, which give a 15:1 ratio in F_2, aabb being the only genotype to give a unique phenotype where the commoner phenotype has the greater expression of the character and AABB being the genotype with the unique phenotype where the commoner class has the lesser expression of the character. In the former case AABB, AABb, AaBB, AaBb, aaBB, aaBb, AAbb and Aabb must have the same phenotype from which it follows that

$$d_a + d_b + i = d_a + h_b + j_a = h_a + d_b + j_b = h_a + h_b + l$$
$$= -d_a + d_b - i = -d_a + h_b - j_a = d_a - d_b - i = h_a - d_b - j_b.$$

These equations are satisfied if $d_a = d_b = h_a = h_b = -i = -j_a = -j_b = -l$. The counterpart situation where AABB is unique and aabb, aaAb, Aabb, AaBb, AAbb, AABb, aaBB and AaBB are alike arises where

$$d_a = d_b = -h_a = -h_b = i = -j_a = -j_b = l.$$

So we see that duplicate interaction arises when all the parameters have the same magnitude, and the two j's are negative while i and l have the opposite sign to the h's. To abbreviate, the condition for complementary action is that

$$d_a = d_b = \pm h_a = \pm h_b = \pm i = j_a = j_b = \pm l$$

while the condition for duplicate interaction is similarly

$$d_a = d_b = \pm h_a = \pm h_b = \mp i = -j_a = -j_b = \mp l.$$

The value of this approach is that we can now generalize the notion of complimentary and duplicate action. For example if we write

$$\theta(d_a = d_b = \pm h_a = \pm h_b) = \pm i = j_a = j_b = \pm l$$

we have no interaction when $i = j = l = 0$ i.e. $\theta = 0$, full complemenary interaction when $\theta = 1$, partial complementary interaction when the interaction parameters are all equal but less than the d's and h's i.e. $0 < \theta < 1$ and over or super-complementary interaction when $\theta > 1$. Furthermore when $\theta = -1$, we have full duplicate interaction, when $0 > \theta > -1$ partial duplicate interaction, and when $-1 > \theta$ over or super-duplicate interaction. We shall see later how this generalization can be put to use. Other more complicated generalizations about interaction are, of course, also possible although none have yet been developed for use in practice.

One last point remains to be made about the classical interactions. An F_2 giving a 9:3:3:1 ratio was regarded classically as showing no interaction. In point of fact a 9:3:3:1 or one of its simple derivatives is obtained whenever $d_a = \pm h_a$, $d_b = \pm h_b$ and $\pm i = j_a = j_b = \pm l$. Thus the ratio does not necessarily indicate an absence of interaction in our sense, but again implies its own limitations in the relations among the interaction parameters.

20. Interaction as displayed by means

A cross producing an F_1 heterozygous for two gene pairs can be made in two ways. The increasing alleles may occur together in one of the true-breeding parents and the decreasing alleles in the other, the cross thus being AABB × aabb, and the genes being said to be associated. Or each parent might carry the increasing allele of one gene and the decreasing allele of the other, the cross thus being AAbb × aaBB, and the genes being said to be dispersed. With association of the genes the parental phenotypes will be $m + d_a + d_b + i$, and $m - d_a - d_b + i$, while with dispersion the phenotypes will be $m + d_a - d_b - i$ and $m - d_a + d_b - i$. The F_1 will have the same phenotype, $m + h_a + h_b + l$, no matter from which cross it is raised. Furthermore, in the absence of linkage, so will the F_2, whose mean can be shown by combining the classes of Table 33 in the F_2 proportions, to be $m + \frac{1}{2}h_a + \frac{1}{2}h_b + \frac{1}{4}l$. It will be observed that again the coefficient of l is the product of the coefficient of the two h's, so illustrating in a new context the general rule for finding the coefficient of an interaction parameter.

Turning to the back-crosses however, the results from the associated and dispersed crosses again differ. With the associated cross, the back-cross to AABB will yield the four genotypes, AABB, AABb, AaBB and AaBb in equal frequencies and the mean will thus be $m + \frac{1}{2}d_a + \frac{1}{2}d_b + \frac{1}{2}h_a + \frac{1}{2}h_b + \frac{1}{4}i + \frac{1}{4}j_a + \frac{1}{4}j_b + \frac{1}{4}l$. Similarly the mean of the back-cross to aabb will be $m - \frac{1}{2}d_a - \frac{1}{2}d_b + \frac{1}{2}h_a + \frac{1}{2}h_b + \frac{1}{4}i - \frac{1}{4}j_a - \frac{1}{4}j_b + \frac{1}{4}l$. With the dispersed cross on the other hand the means of the two back-crosses will be

$$m + \tfrac{1}{2}d_a - \tfrac{1}{2}d_b + \tfrac{1}{2}h_a + \tfrac{1}{2}h_b - \tfrac{1}{4}i + \tfrac{1}{4}j_a - \tfrac{1}{4}j_b + \tfrac{1}{4}l$$

and

$$m - \tfrac{1}{2}d_a + \tfrac{1}{2}d_b + \tfrac{1}{2}h_a + \tfrac{1}{2}h_b - \tfrac{1}{4}i - \tfrac{1}{4}j_a + \tfrac{1}{4}j_b + \tfrac{1}{4}l$$

respectively.

These results are collected together in Table 34. Before, however, they can be used in the analysis of experimental data they must be generalized to cover the case of more than two genes. As we saw in Section 8, the d's of the different genes must tend to balance one another out where the genes are dispersed, so leading us to define $[d]$ as the sum of the d's taking sign into account where some genes are associated in the parents while others are dispersed. We also defined $[h]$ as the sum of the h's of the individual genes taking sign into account, although here the sign of h does not depend on gene association nor dispersion but on the direction of the dominance itself. In the same way with k gene differences there will prospectively be $\frac{1}{2}k(k-1)$ digenic interactions of types i and l and $k(k-1)$ digenic interactions of type j, since each pair of genes prospectively yields two j interactions, j_{ab} and j_{ba}. We must therefore define $[i]$, $[j]$ and $[l]$ as being respectively the sums of the $\frac{1}{2}k(k-1)$ i and l interactions and of the $k(k-1)$ j interactions, taking sign into account. Now with l, as with h, sign will depend solely on the direction of the interac-

TABLE 34.

Interactions in the means of families of a digenic cross

	m	d_a	d_b	h_a	h_b	i_{ab}	j_{ab}	j_{ba}	l_{ab}
Associated	AABB × aabb								
$\overline{P}_1$	1	1	1			1			
$\overline{P}_2$	1	−1	−1			1			
$\overline{F}_1$	1			1	1				1
$\overline{F}_2 = \overline{S}_3$	1			$\frac{1}{2}$	$\frac{1}{2}$				$\frac{1}{4}$
$\overline{F}_3$	1			$\frac{1}{4}$	$\frac{1}{4}$				$\frac{1}{16}$
$\overline{B}_1$	1	$\frac{1}{2}$	$\frac{1}{2}$	$\frac{1}{2}$	$\frac{1}{2}$	$\frac{1}{4}$	$\frac{1}{4}$	$\frac{1}{4}$	$\frac{1}{4}$
$\overline{B}_2$	1	$-\frac{1}{2}$	$-\frac{1}{2}$	$\frac{1}{2}$	$\frac{1}{2}$	$\frac{1}{4}$	$-\frac{1}{4}$	$-\frac{1}{4}$	$\frac{1}{4}$
Dispersed	AAbb × aaBB								
$\overline{P}_1$	1	1	−1			−1			
$\overline{P}_2$	1	−1	1			−1			
$\overline{B}_1$	1	$\frac{1}{2}$	$-\frac{1}{2}$	$\frac{1}{2}$	$\frac{1}{2}$	$-\frac{1}{4}$	$\frac{1}{4}$	$-\frac{1}{4}$	$\frac{1}{4}$
$\overline{B}_2$	1	$-\frac{1}{2}$	$\frac{1}{2}$	$\frac{1}{2}$	$\frac{1}{2}$	$-\frac{1}{4}$	$-\frac{1}{4}$	$\frac{1}{4}$	$\frac{1}{4}$

tion, being positive when, for example, the two h's and l yielded by two genes are in the same direction, and negative when l is in the opposite direction to the h's. With i and j interaction, however, not only does the direction of the interaction itself enter in, but also whether the two genes in question are associated or dispersed in the parents, as indeed we can see from Table 34. The i yielded by two genes will be in one direction when the genes are associated but in the other when they are dispersed, whereas if they are intrinsically in the same direction the two j's will reinforce one another when the genes are associated but will tend to cancel one another out when the genes are dispersed. The algebraic relations of i and j to the proportions of the k genes which are associated and dispersed is somewhat complex (see M and J) and need not be detailed here. It is sufficient for us to note that neither $[i]$ nor $[j]$ need be 0 in a given cross even where $[d] = 0$ as a result of partial dispersion of the genes. As with $[d]$ and $[h]$, however, $[i] = 0$ does not necessarily imply that all the individual i's are 0, although $[i] \neq 0$ must imply that at least some of the i's are not 0. The same is of course true of $[j]$ and $[l]$.

We can see from Table 34, but using the generalized forms for $[d]$, $[h]$ and their interactions, which take into account the effects of association and dispersion as well as the direction of the interaction

$$\overline{P}_1 = m + [d] + [i]$$

$$\begin{aligned}
\bar{P}_2 &= m - [d] + [i] \\
\bar{F}_1 &= m + [h] + [l] \\
\bar{F}_2 &= m + \tfrac{1}{2}[h] + \tfrac{1}{4}[l] \\
\bar{B}_1 &= m + \tfrac{1}{2}[d] + \tfrac{1}{2}[h] + \tfrac{1}{4}[i] + \tfrac{1}{4}[j] + \tfrac{1}{4}[l] \\
\bar{B}_2 &= m - \tfrac{1}{2}[d] + \tfrac{1}{2}[h] + \tfrac{1}{4}[i] - \tfrac{1}{4}[j] + \tfrac{1}{4}[l].
\end{aligned}$$

Six parameters are involved in these expressions and six means are available for their estimation. We can therefore arrive at perfect fit estimates of the six parameters, thus

$$\begin{aligned}
m &= \tfrac{1}{2}\bar{P}_1 + \tfrac{1}{2}\bar{P}_2 + 4\bar{F}_2 - 2\bar{B}_1 - 2\bar{B}_2 \\
[d] &= \tfrac{1}{2}\bar{P}_1 - \tfrac{1}{2}\bar{P}_2 \\
[h] &= 6\bar{B}_1 + 6\bar{B}_2 - 8\bar{F}_2 - \bar{F}_1 - 1\tfrac{1}{2}\bar{P}_1 - 1\tfrac{1}{2}\bar{P}_2 \\
[i] &= 2\bar{B}_1 + 2\bar{B}_2 - 4\bar{F}_2 \\
[j] &= 2\bar{B}_1 - \bar{P}_1 - 2\bar{B}_2 + \bar{P}_2 \\
[l] &= \bar{P}_1 + \bar{P}_2 + 2\bar{F}_1 + 4\bar{F}_2 - 4\bar{B}_1 - 4\bar{B}_2.
\end{aligned}$$

The standard errors of these estimates can be found in the usual way. Thus, for example,

$$V_{[d]} = \tfrac{1}{4}V_{\bar{P}_1} + \tfrac{1}{4}V_{\bar{P}_2} \quad \text{and} \quad s_{[d]} = \sqrt{V_{[d]}}.$$

The significance of $[d]$ can then be tested by calculating

$$t = [d]/s_{[d]}.$$

Finding $[i]$, $[j]$ or $[l]$ significant in such tests is obviously equivalent to finding significant deviations from zero in the scaling tests; but it has the additional advantage of yielding estimates of the parameters and therefore of identifying the type or types of interactions responsible for the departure from the simple additive-dominance situation. We should note that the 3 degrees of freedom, from which is derived the $\chi^2_{[3]}$ testing the goodness of fit of the model in the joint scaling test described on pages 37–40, are now being used for estimating the three interaction parameters. No test of goodness of fit is therefore possible of the new model incorporating the three types of digenic interaction: indeed as we have seen it is a perfect fit estimation. More generations such as F_3 or second backcrosses must be included if sufficient equations are to be available to provide a test of goodness of fit. If in such a case the model involving digenic interactions proves to be inadequate to account for the results, we should have to consider the possibility of trigenic interaction or some other further complicating factor but this is beyond the scope of our present treatment.

We may illustrate the procedure of estimation in a simple case by reference once more to data from the cross between varieties 72 and 22 of *Nicotiana rustica* for plant height six weeks after planting in the field which was analysed in Chapter 3. The *C* scaling test and the joint scaling test when applied to these data were highly significant (Table 8). The simple additive-dominance model is clearly inadequate. Furthermore, attempts to find an alternative scale on which this model would be adequate failed. If we wish to analyse these data further we must, therefore, allow for the presence of non-allelic interaction (or epistasis as it is sometimes called) in any model we attempt to fit.

Using the perfect fit formulae we can estimate the three interaction components, $[i]$, $[j]$ and $[l]$ in addition to m, $[d]$ and $[h]$. As we have already seen

$$[d] = \tfrac{1}{2}\bar{P}_1 - \tfrac{1}{2}\bar{P}_2.$$

On substituting the appropriate family means from Table 8, this gives

$$\begin{aligned}[d] &= \tfrac{1}{2}(80.40 - 65.47)\\ &= 7.46.\end{aligned}$$

Similarly,

$$\begin{aligned}s_{[d]} &= \sqrt{V_{[d]}} = \sqrt{[\tfrac{1}{4}(1.936)^2 + \tfrac{1}{4}(1.726)^2]}\\ &= \sqrt{1.680} = \pm 1.296.\end{aligned}$$

The t for 38 df for testing the significance of $[d]$ is therefore

$$t_{[38]} = \frac{7.46}{1.30} = 5.74$$

which has a probability of $P < 0.001$.

These results along with those for the other five components are summarized in Table 35. Five of the estimates are significant, including the two interaction components $[i]$ and $[l]$. The significance of these two interactions components confirms the earlier conclusions from the scaling tests. Now because we have estimated six components from six observed means we have no test of the adequacy of the present model. Normally we would have to raise further generations to provide such a test. Since, however, the estimate of one of the interaction components,

TABLE 35.

Estimates of the additive, dominance and digenic interaction components of means for plant height in the cross between varieties 72 and 22 of *Nicotiana rustica*

Component	Perfect fit estimate	P	Five component estimate	P
m	92.93 ± 4.76	<0.001	93.50 ± 4.60	<0.001
[d]	7.46 ± 1.30	<0.001	8.64 ± 0.99	<0.001
[h]	−28.64 ± 12.21	0.05 − 0.01	−30.27 ± 12.13	0.05 − 0.01
[i]	−19.99 ± 4.61	<0.001	−20.43 ± 4.60	<0.001
[j]	5.68 ± 4.03	>0.05	——	–
[l]	21.71 ± 7.91	0.01 − 0.001	22.86 ± 7.88	0.01 − 0.001
		$\chi^2_{[1]}$	1.99	0.20 − 0.10

[j], does not differ significantly from zero it would appear that a model in which it was omitted would be adequate for these data.

Fitting a five parameter model by omitting [j] would allow us to test the goodness of fit of the model by means of a χ^2 with one df, and at the same time improve the precision with which the remaining parameters were estimated. Estimating the five components of this model proceeds exactly as for the simple additive-dominance model in the joint scaling test (Chapter 3, Section 9). It leads to the estimates on the right-hand side of Table 35. As expected the five parameter model is adequate, the $\chi^2_{[1]}$ testing its goodness of fit being non-significant. There is also a marginal improvement in the precision with which we have estimated the five components, as shown by their lower standard errors.

Since the model is adequate we can conclude that trigenic interactions and similar complex factors are not making a significant contribution to the differences among the generation means. We can interpret the data, therefore, in terms of the additive, dominance and digenic non-allelic interaction components of the gene action. The h increments of the majority of individual loci must be negative while the l increments of the majority of pairs of loci must be positive. The non-allelic interaction is, therefore, mainly of the duplicate kind.

Before leaving the effects of non-allelic interaction on means we must note the contribution it can make to heterosis. Heterosis will be observed when $\bar{F}_1 > \bar{P}_1$, where $\bar{P}_1$ is taken as the parent with the greater expression of the character. As we have seen earlier, in the absence of interaction $\bar{F}_1 > \bar{P}_1$ requires that [h] > [d], and this in turn requires that one or both of two conditions be satisfied, namely

(i) $h > d$ for some or all of the genes; that is there must be overdominance at some or all loci.

(ii) $[d] < Sd$; that is there must be dispersion of the genes between the parents, the value of $[d]$ being thus reduced by the balancing effects of the genes of opposite effect in each parent, whereby $[h]$ may exceed $[d]$ although each h is no larger and may even be smaller than its corresponding d.

These two conditions cannot be distinguished from means alone, although second degree statistics allow the distinction to be made. At the same time it is a distinction of great practical importance, since wherever heterosis depends on overdominance the maximum expression of the character, for example yield in a crop plant, can be achieved only by a hybrid breeding programme producing F_1's for commercial use. Where, however, heterosis is due to dispersion of the genes, it is in principle always possible to produce a true breeding line expressing the character to at least as high a degree as the F_1, although of course this may involve the breakage of linkages between the dispersed genes.

Now where digenic interaction is displayed the requirement for $\bar{F}_1 > \bar{P}_1$ becomes $[h] + [l] > [d] + [i]$. This relation clearly offers a number of possibilities for the production of heterosis. Two effects, reinforcing the relations by which heterosis may arise in the absence of interaction, are however of special importance, namely

(i) That the h's and their associated l's are entirely or at any rate preponderantly of the same sign, which of course is a feature of complementary gene action.

(ii) Dispersion of the interacting genes between the parents, so that although, as is required by complementary interaction, the sign of the individual i's is the same as that of the h's, $[i]$ will take a negative sign in the parents.

The first relation will raise the value of $[h] + [l]$, the expression of the character in F_1. The second will limit the increase in value of $[d] + [i]$, and may even diminish it relative to $[d]$.

Thus complementary interaction can increase the expression of heterosis whether it be due to over-dominance or gene dispersion. It is thus not surprising that wherever the data permit the analysis to be made, non-allelic interaction, presumably of the complementary type, has been found to be a common accompaniment of heterosis. These effects of digenic interaction on heterosis are illustrated in Fig. 11.

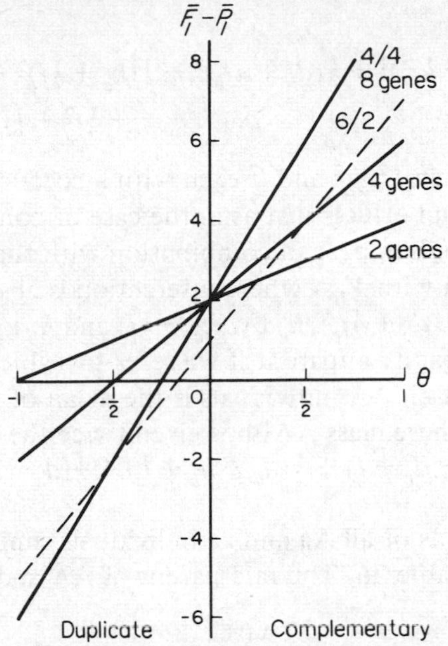

Fig. 11. Heterosis, measured by the excess of the F_1 mean over that of the better parent ($\bar{F}_1 - \bar{P}_1$), in relation to non-allelic interaction, measured by θ. Solid lines show the relationship where 2, 4, and 8 gene pairs are respectively involved with maximum dispersal, i.e. 1 increasing allele in each parent for 2 gene pairs (1/1), 2 in each parent for 4 gene pairs (2/2) and 4 in each parent for 8 gene pairs (4/4). The broken line shows the relationship for 8 genes where 6 increasing alleles are in one parent and 2 in the other (6/2). Note that in all cases, except that of 2 gene pairs, the sign of the heterosis is reversed where duplicate type interaction of sufficient strength is operating. The diagram assumes that all d's are equal to one another and to all h's, with all i's = all l's = θd.

21. Variances and covariances

Although the means of the parent lines and the F_1 reflect the effects of non-allelic interaction, their variances are unaffected because being genetically uniform their variation is entirely non-heritable. Turning to F_2 we can find its variance by squaring the phenotype of each of the nine genotypes as set out in Table 33, multiplying by the frequencies with which they appear in F_2, summing and subtracting the square of the mean phenotype, thus

$$V_{1F2} = \tfrac{1}{16}(d_a + d_b + i)^2 + \tfrac{1}{8}(h_a + d_b + j_b)^2 \ldots + \tfrac{1}{4}(h_a + h_b + l)^2$$
$$\ldots + \tfrac{1}{16}(-d_a - d_b + i)^2 - (\tfrac{1}{2}h_a + \tfrac{1}{2}h_b + \tfrac{1}{4}l)^2$$

which reduces to

$$V_{1F2} = \tfrac{1}{2}(d_a + \tfrac{1}{2}j_a)^2 + \tfrac{1}{2}(d_b + \tfrac{1}{2}j_b)^2 + \tfrac{1}{4}(h_a + \tfrac{1}{2}l)^2 + \tfrac{1}{4}(h_b + \tfrac{1}{2}l)^2 + \tfrac{1}{4}i^2 + \tfrac{1}{8}j_a^{\,2} + \tfrac{1}{8}j_b^{\,2} + \tfrac{1}{16}l^2.$$

Terms appear in i^2, $j_a^{\,2}$, $j_b^{\,2}$ and l^2 each with a coefficient the product of its two relevant main effects just as in the case of contribution to means, but in addition the j's appear in combination with the d's and l with the h's. By comparison with V_{1F2} where interaction is absent, d_a is replaced by $(d_a + \tfrac{1}{2}j_a)$, d_b by $(d_b + \tfrac{1}{2}j_b)$, h_a by $(h_a + \tfrac{1}{2}l)$ and h_b by $(h_b + \tfrac{1}{2}l)$. The reason for this is readily apparent if we refer to Table 33. In an F_2 the mean expression of all AA individuals is the mean of the classes AABB, AABb and aaBB where class AABb is given twice the weight of the other two, i.e. it is $\tfrac{1}{4}(d_a + d_b + i) + \tfrac{1}{2}(d_a + h_b + j_a) + \tfrac{1}{4}(d_a - d_b - i) = d_a + \tfrac{1}{2}j_a + \tfrac{1}{2}h_b$.

Finding the means of all Aa and aa individuals similarly we obtain the results set out in Table 36. The mid-parent of AA and aa homozygotes is

TABLE 36.

Mean phenotypes of AA, Aa and aa classes in F_2 and F_3, expressed as deviations from the mid-parent, m

Class	Mean	Deviation	Mean	Deviation
AA	$d_a + \tfrac{1}{2}j_a + \tfrac{1}{2}h_b$	$d_a + \tfrac{1}{2}j_a$	$d_a + \tfrac{1}{4}j_a + \tfrac{1}{4}h_b$	$d_a + \tfrac{1}{4}j_a$
Aa	$h_a + \tfrac{1}{2}l + \tfrac{1}{2}h_b$	$h_a + \tfrac{1}{2}l$	$h_a + \tfrac{1}{4}l + \tfrac{1}{4}h_b$	$h_a + \tfrac{1}{4}l$
aa	$-d_a - \tfrac{1}{2}j_a + \tfrac{1}{2}h_b$	$-(d_a + \tfrac{1}{2}j_a)$	$-d_a - \tfrac{1}{4}j_a + \tfrac{1}{4}h_b$	$-(d_a + \tfrac{1}{4}j_a)$
$\tfrac{1}{2}(\overline{AA} + \overline{aa})$	$\tfrac{1}{2}h_b$		$\tfrac{1}{4}h_b$	

$\tfrac{1}{2}h_b$ and the deviation from it of AA, aa and Aa are respectively $(d_a + \tfrac{1}{2}j_a)$, $-(d_a + \tfrac{1}{2}j_a)$ and $(h_a + \tfrac{1}{2}l)$. In the case of gene B-b, the corresponding deviations are $(d_b + \tfrac{1}{2}j_b)$, $-(d_b + \tfrac{1}{2}j_b)$ and $(h_b + \tfrac{1}{2}l)$. These deviations replace d_a, $-d_a$, h_a, d_b, $-d_b$, and h_b which obtain in the absence of interaction.

If we pass on to F_3, taking the generation as a whole, the four complete homozygotes (AABB, etc.) each comprise $\tfrac{9}{64}$, the four single heterozygotes (AABb, etc.) each comprise $\tfrac{6}{64}$, and the doubly heterozygous genotype (AaBb) comprises $\tfrac{4}{64}$ of the individuals. The means of all AA, Aa and aa individuals are thus $d_a + \tfrac{1}{4}j_a + \tfrac{1}{4}h_b$, $h_a + \tfrac{1}{4}l + \tfrac{1}{4}h_b$ and $-(d_a + \tfrac{1}{4}j_a) + \tfrac{1}{4}h_b$ giving deviations of $(d_a + \tfrac{1}{4}j_a)$, $(h_a + \tfrac{1}{4}l)$ and $-(d_a + \tfrac{1}{4}j_a)$. It is not surprising therefore to find that the total variance of the F_3 generation is

$$V_{F3} = V_{1F3} + V_{2F3} = \tfrac{3}{4}(d_a + \tfrac{1}{4}j_a)^2 + \tfrac{3}{4}(d_b + \tfrac{1}{4}j_b)^2 + \tfrac{3}{16}(h_a + \tfrac{1}{4}l)^2 + \tfrac{3}{16}(h_b + \tfrac{1}{4}l)^2 + \tfrac{9}{16}i^2 + \tfrac{9}{64}j_a{}^2 + \tfrac{9}{64}j_b{}^2 + \tfrac{9}{256}l^2$$

the coefficients of the terms in i^2, j^2 and l^2 being once again the products of the coefficients of the relevant main effects. If we proceed further to find V_{1F3} and V_{2F3} we again find terms in $(d + \frac{1}{4}j)^2$ and $(h + \frac{1}{4}l)^2$, thus

$$V_{1F3} = \tfrac{1}{2}(d_a + \tfrac{1}{4}j_a)^2 + \tfrac{1}{2}(d_b + \tfrac{1}{4}j_b)^2 + \tfrac{1}{16}(h_a + \tfrac{1}{4}l)^2 + \tfrac{1}{16}(h_b + \tfrac{1}{4}l)^2 + \tfrac{1}{4}i^2 + \tfrac{1}{32}j_a{}^2 + \tfrac{1}{32}j_b{}^2 + \tfrac{1}{256}l^2$$

$$V_{2F3} = \tfrac{1}{4}(d_a + \tfrac{1}{4}j_a)^2 + \tfrac{1}{4}(d_b + \tfrac{1}{4}j_b)^2 + \tfrac{1}{8}(h_a + \tfrac{1}{4}l)^2 + \tfrac{1}{8}(h_b + \tfrac{1}{4}l)^2 + \tfrac{5}{16}i^2 + \tfrac{7}{64}j_a{}^2 + \tfrac{7}{64}j_b{}^2 + \tfrac{1}{32}l^2.$$

The coefficients of the interaction terms are again the products of the coefficients of the relevant main effects in V_{1F3}, but not in V_{2F3}. Indeed since $V_{1F3} + V_{2F3} = V_{F3}$ the product rule cannot apply to V_{2F3} if it applies to V_{1F3} and V_{F3}.

The covariance of F_2 parents and F_3 means is

$$W_{1F23} = \tfrac{1}{2}(d_a + \tfrac{1}{2}j_a)(d_a + \tfrac{1}{4}j_a) + \tfrac{1}{2}(d_b + \tfrac{1}{2}j_b)(d_b + \tfrac{1}{4}j_b) + \tfrac{1}{8}(h_a + \tfrac{1}{2}l)(h_a + \tfrac{1}{4}l) + \tfrac{1}{8}(h_b + \tfrac{1}{2}l)(h_b + \tfrac{1}{4}l) + \tfrac{1}{4}i^2 + \tfrac{1}{16}j_a{}^2 + \tfrac{1}{16}j_b{}^2 + \tfrac{1}{64}l^2$$

the product rule applying once again to the interaction coefficients. The expressions involving d_a, d_b, h_a and h_b are, not surprisingly, the geometric means of their counterparts in V_{1F2} and V_{1F3}. We can proceed to find similar expressions for V_{1F4}, V_{2F4}, and V_{3F4}, which all include terms in $(d_a + \frac{1}{8}j_a)$, $(d_b + \frac{1}{8}j_b)$, $(h_a + \frac{1}{8}l)$ and $(h_b + \frac{1}{8}l)$. The covariances W_{1F34} and W_{2F34} similarly include terms in $(d_a + \frac{1}{4}j_a)(d_a + \frac{1}{8}j_a)$ etc.

The various expressions relate only to the effects of two gene pairs, A-a and B-b. They require generalization in two ways. In the first place in so far as further genes C-c, D-d, etc. are involved, their digenic interaction with A-a will be covered for F_2 if for $(d_a + \frac{1}{2}j_a)$ we substitute $(d_a + \frac{1}{2}\mathrm{S}j_a)$ and for $(h_a + \frac{1}{2}l)$ we substitute $(h_a + \frac{1}{2}\mathrm{S}\,l_a)$ where $\mathrm{S}\,j_a$ is the sum of j_{ab}, j_{ac} etc. and $\mathrm{S}l_a$ is the sum of l_{ab}, l_{ac} etc. The further interactions with B-b are covered by the corresponding substitutions of $\mathrm{S}j_b$ and $\mathrm{S}l_b$ for j_b and l_b. The second stage is the generalization of the expressions to cover all genes showing digenic interactions by writing $V_{1F2} = \frac{1}{2}D + \frac{1}{4}H + I$

where $D = \mathrm{S}(d_a + \frac{1}{2}\mathrm{S}j_a)^2$, $H = \mathrm{S}(h_a + \frac{1}{2}\mathrm{S}l_a)^2$

and $I = \frac{1}{4}\mathrm{S}(i^2) + \frac{1}{8}\mathrm{S}(j^2) + \frac{1}{16}\mathrm{S}(l^2)$.

Similarly $V_{1F3} = \frac{1}{2}D + \frac{1}{16}H + I$ and $V_{2F3} = \frac{1}{4}D + \frac{1}{8}H + I$

where

$$D = S(d_a + \tfrac{1}{4}Sj_a)^2, H = S(h_a + \tfrac{1}{4}Sl_a)^2 \text{ and}$$
$$I = \tfrac{1}{4}S(i^2) + \tfrac{1}{32}S(j^2) + \tfrac{1}{265}S(l^2) \text{ in } V_{1F3}$$

and $I = \frac{5}{16}S(i^2) + \frac{7}{64}S(j^2) + \frac{1}{32}S(l^2)$ in V_{2F3}. Each expression should of course have an appropriate E attached to it to accommodate non-heritable variation.

Leaving aside the term I for the moment, these expressions are the same as already found for V_{1F2}, etc., in the absence of interaction, whose effects are accommodated by changing the definition of D from $S(d_a^2)$ to $S(d_a + wSj_a)^2$ and that of H from $S(h_a^2)$ to $S(h_a + wSl_a)^2$ the coefficient w changing with the generation, being $\frac{1}{2}$ in F_2, $\frac{1}{4}$ in F_3, $\frac{1}{8}$ in F_4 and so on. The evidence of non-allelic interaction at least of the j and l types is thus provided by a test of homogeneity of D and H over generations. The term I is a distraction in such a test. It too is inhomogenous over generations but it is also inhomogeous within generations. Short of the cumbersome and demanding estimation and testing of $S(i^2)$, $S(j^2)$ and $S(l^2)$ as individual parameters it is not easy to deal with the inhomogeneity of I without assuming some relation between $S(i^2)$, $S(j^2)$ and $S(l^2)$. There has as yet been insufficient study of interaction to provide any basis for the handling of I, and indeed beyond demonstrating that interactions are exerting their effects in distorting the second degree statistics from which we estimate D and H, experimental studies have provided little information about the way their consequences are revealed by these second degree statistics.

We are assuming that genes A-a and B-b are unlinked. It therefore makes no difference to the variances and covariances of F_2, F_3, etc. and indeed to S_2, S_3 and other generations derived directly from the initial cross, whether this was AABB × aabb or AAbb × aaBB. This is not true, however, of the statistics obtained from families obtained by back-crossing to the parents. Just as we have seen to be the case with the means of the back-cross families, their variances differ according to whether the genes were associated or dispersed in the parents of the cross. Again just as in the case of the means (Table 34) these differences appear in the signs the interaction parameters take in the various terms of the variances. This is well illustrated by the summed variances of the two back-crosses which is

$$V_{B1} + V_{B2} = \tfrac{1}{2}(d_a + \tfrac{1}{2}j_a \pm \tfrac{1}{2}j_b)^2 + \tfrac{1}{2}(d_b \pm \tfrac{1}{2}j_a + \tfrac{1}{2}j_b)^2$$
$$+ \tfrac{1}{2}(h_a \pm \tfrac{1}{2}i + \tfrac{1}{2}l)^2 + \tfrac{1}{2}(h_b \pm \tfrac{1}{2}i + \tfrac{1}{2}l)^2 + \tfrac{1}{8}(i \pm l)^2 + \tfrac{1}{8}(j_a \pm j_b)^2$$

where in the case of a double sign the upper one applies where the genes were associated in the parents of the cross (AABB × aabb) and the lower where they were dispersed (AAbb × aaBB). It will be seen too that the summed variances of the back-crosses differ from the variances of F_2, F_3 etc. not only in their dependence on the distribution of the genes between the parents but also in the interaction items which are associated with d and h in the relevant terms. Thus in $V_{B1} + V_{B2}$ both j interactions appear with appropriate signs, in both the d_a and d_b terms, and i appears as well as l in the h terms. Furthermore in the purely interaction terms themselves, i, j and l do not contribute separately, but i is always joined with l and j_a with j_b. Once again D and H as they appear in $V_{B1} + V_{B2}$ will be inhomogeneous with D and H as they appear in F_2 etc., as will also the I term. $V_{B1} + V_{B2}$ can thus be brought directly into the test of second degree statistics for the effects of non-allelic interaction.

With back-crosses the effects of interaction in inflating or reducing the variances will depend, at least in some measure, on the association or dispersion of the genes in the parents. In F_2 and its derived generations this is not the case: inflation or reduction of the variances depends only on the direction and nature of the interaction, that is on the intrinsic signs of the interaction parameters themselves. In

$$V_{1F2} = \tfrac{1}{2}(d_a + \tfrac{1}{2}j_a)^2 + \tfrac{1}{2}(d_b + \tfrac{1}{2}j_b)^2 + \tfrac{1}{4}(h_a + \tfrac{1}{2}l)^2 + \tfrac{1}{4}(h_b + \tfrac{1}{2}l)^2$$
$$+ \tfrac{1}{4}i^2 + \tfrac{1}{8}j_a^2 + \tfrac{1}{8}j_b^2 + \tfrac{1}{16}l^2$$

i will always tend to increase the variance, but j will tend to increase it when positive and generally to decrease it when negative. Equally, l will tend to increase the variance when of the same sign as h but will generally decrease it when of opposite sign. Thus in complementary type interaction where, as we saw in Section 19, j must always be positive and i and l the same sign as h, the interaction must always inflate the value of V_{1F2}, to an extent depending on θ. It will equally inflate the variances in F_3, F_4, etc. although to varying degrees depending especially on w, the coefficient of j and l in the terms contributing to D and H. Equally in duplicate type interaction, where j is always negative while i and l are of opposite sign to h, the interaction will tend to reduce the variances of F_2 and its derived generations, again to varying degrees depending on the value of θ, until θ attains the critical ratio where the depressing effect of j on $\frac{1}{2}(d + \frac{1}{2}j)^2$ is offset by the increase due to the $\frac{1}{8}j^2$ term, and the effect of l on $\frac{1}{4}(h + \frac{1}{2}l)^2$ is offset by the term $\frac{1}{16}l^2$. This ratio is reached in F_2 when $\theta = -1.6$ where only two gene pairs are involved in the interaction, but because of the cumulative effects of the interaction in the D and H components it is attained in values of θ nearer to zero as the number of genes involved in the system rises. Thus complementary interactions always tend to raise the variances of F_2 and its derived generations, but duplicate interaction tends to reduce these variances at least when θ is not unduly large (see Fig. 12).

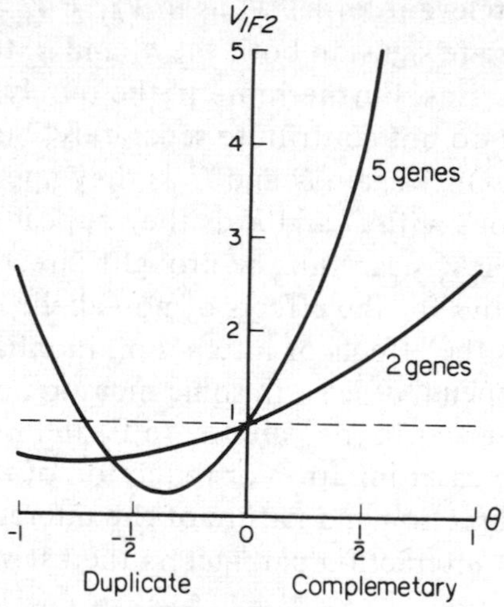

Fig. 12. The effect on V_{IF2} of complementary and duplicate type interaction, measured by θ, in the cases of 2 and 5 segregating gene pairs. In each of the two cases all d = all h, and all i = all j = all $l = \theta d$. In both the 2 and 5 gene cases the values of V_{IF2} are scaled to be 1 when there is no interaction ($\theta = 0$).

22. Correlated gene distributions: linkage

The second cause of non-independence of the effects of the various genes on the phenotype is the correlation of their distributions among the individuals of the families, groups or generations under observation. In the generations derived from a cross between true-breeding parents, the primary cause of correlated distributions of the genes is linkage, to whose consequences must we turn first.

Consider first the consequences of linkage for the mean expression of the character. Now, of itself, linkage does not affect the frequencies with which the alleles of each gene pair are recovered in segregating generations: it only leads to particular combinations of the alleles of different gene pairs appearing with frequencies other than those expected on the basis of independence. In the absence of non-allelic interaction the increments added to the phenotype by the various gene pairs are additive and the average effect of a gene on the phenotype will be the same, apart from sampling variation, no matter what its linkage relations may be: the relative frequencies of particular combinations in which the alleles occur with other non-allelic genes will have no effect, because every one which is over-common will be balanced by another which is correspondingly rare. Linkage therefore can of itself have no effect on the mean measurements of segregating families provided that no non-allelic interaction is present; and indeed the same will be true of any correlation of gene distribution whatever its cause, provided it does not alter the frequencies with which the combinations of allelic genes are recovered.

Thus linkage will not vitiate the use of the scaling tests for detecting departures from the assumption of no non-allelic interaction. At the same time however, where non-allelic interaction is indeed present, linkage will affect the contribution of this interaction to the mean expressions of segregating generations: since it determines the relative frequencies with which different combinations of non-allelic genes appear, it will determine the frequencies with which the different types of interaction, i, j and l, arise. This is, however, a complex subject (see M and J, Section 18), which we will not pursue beyond noting that where the frequency of recombination between A-a and B-b is p, and $q = 1 - p$

$$\bar{F}_2 = m + \tfrac{1}{2}(h_a + h_b) + \tfrac{1}{2}(1 - 2p)i + \tfrac{1}{2}(1 - 2pq)l$$

$$\bar{B}_1 = m + \tfrac{1}{2}(d_a \pm d_b) + \tfrac{1}{2}(h_a + h_b) \pm \tfrac{1}{2}(1 - 2p)i + \tfrac{1}{2}p(j_a \pm j_b) + \tfrac{1}{2}(1 - p)l$$

$$\bar{B}_2 = m - \tfrac{1}{2}(d_a \pm d_b) + \tfrac{1}{2}(h_a + h_b) \pm \tfrac{1}{2}(1 - 2p)i - \tfrac{1}{2}p(j_a \pm j_b) + \tfrac{1}{2}(1 - p)l$$

where in the case of a double sign the upper refers to coupled genes and the lower to repulsion, as association and dispersion may properly be styled in the case of linkage.

Turning to second degree statistics, consider the simplest case of two genes A-a and B-b. Where the frequency of recombination between them is again p and they are in coupling the ten genotypes are expected in F_2 with the frequencies shown in Table 37, which also shows the phenotypic deviations from m and the mean phenotypes of the corresponding F_3 families, both on the assumption of no non-allelic interaction. The mean of F_2 is $\tfrac{1}{2}(h_a + h_b)$ being unaffected by linkage. The heritable variance of F_2 is found as

TABLE 37.

Frequencies, F_2 phenotypes and F_3 mean phenotypes of the ten genotypic classes in an F_2 for two coupled genes. In each cell the uppermost entry is the frequency, the middle entry is the F_2 phenotype (expressed as a deviation from the mid-parent) and the lowest is the F_3 mean. All frequencies should be divided by four. C indicates coupling and R repulsion double heterozygotes

	AA	Aa	aa
BB	q^2 $d_a + d_b$ $d_a + d_b$	$2pq$ $h_a + d_b$ $\frac{1}{2}h_a + d_b$	p^2 $-d_a + d_b$ $-d_a + d_b$
Bb	$2pq$ $d_a + h_b$ $d_a + \frac{1}{2}h_b$	C$2q^2$ $h_a + h_b$ $\frac{1}{2}h_a + \frac{1}{2}h_b$ $2p^2$ $h_a + h_b$ $\frac{1}{2}h_a + \frac{1}{2}h_b$ R	$2pq$ $-d_a + h_b$ $-d_a + \frac{1}{2}h_b$
bb	p^2 $d_a - d_b$ $d_a - d_b$	$2pq$ $h_a - d_b$ $\frac{1}{2}h_a - d_b$	q^2 $-d_a - d_b$ $-d_a - d_b$

$$\begin{aligned} V_{1F2} &= \tfrac{1}{4}[q^2(d_a + d_b)^2 + 2pq(h_a + d_b)^2 \ldots \\ &\qquad + q^2(-d_a - d_b)^2] - [\tfrac{1}{2}(h_a + h_b)]^2 \\ &= \tfrac{1}{2}[d_a^2 + d_b^2 + 2(1-2p)d_a d_b] \\ &\qquad + \tfrac{1}{4}[h_a^2 + h_b^2 + 2(1-2p)^2 h_a h_b]. \end{aligned}$$

The two hitherto unfamiliar terms in this expression involve the recombination value, combined in one case with $d_a d_b$ and in the other with $h_a h_b$. With free recombination $p = \frac{1}{2}$, $1 - 2p = 0$ and the new terms vanish to leave the expressions obtained in Section 11. With complete linkage $p = 0$, $1 - 2p = 1$ and, aside from non-heritable variation, $V_{1F2} = \frac{1}{2}(d_a + d_b)^2 + \frac{1}{4}(h_a + h_b)^2$. The two genes are then acting as one. Even where recombination occurs, however, the recombinant genotypes will be rare if p is small, and the genes will effectively act as one except in so far as selection may isolate one of the rare recombinants.

Where the genes are in repulsion the heritable variance of F_2 becomes

$$V_{1F2} = \tfrac{1}{2}[d_a^2 + d_b^2 - 2(1-2p)d_a d_b] + \tfrac{1}{4}[h_a^2 + h_b^2 + 2(1-2p)^2 h_a h_b].$$

The sign of the term in $d_a d_b$ is changed but, as would be expected, that in $h_a h_b$ remains the same. It should be noted, however, that $h_a h_b$ will be positive only if h_a and h_b are reinforcing one another by acting in the same direction. If they are opposing one another in action this term will take a negative sign. Thus reinforcement versus opposition of the h's resembles coupling versus repulsion of the genes in its effects on the signs of the term in p. It should be remembered nevertheless that reinforcement versus opposition is a physiological distinction while coupling versus repulsion is a mechanical one.

If we now write

$$D = d_a^2 + d_b^2 \pm 2(1-2p)\,d_a\,d_b$$

and

$$H = h_a^2 + h_b^2 + 2(1-2p)^2\,h_a\,h_b$$

where the $\pm$ of the term in $d_a\,d_b$ denotes + for coupling and − for repulsion, we can put

$$V_{1F2} = \tfrac{1}{2}D + \tfrac{1}{4}H + E.$$

Furthermore, it is easy to show by reference to Table 37

$$V_{1F3} = \tfrac{1}{2}D + \tfrac{1}{16}H + E$$
$$W_{1F23} = \tfrac{1}{2}D + \tfrac{1}{8}H$$

with the same definitions of D and H.

This revision of the definition of D and H, by comparison with those of Section 12, accommodates the effect of linkage on the variation as expressed in any variance or covariance of rank 1 (indicated by the initial 1 in the subscript of, for example, V_{1F2}). Now just as the mean of F_2 is unaffected by the linkage relations of the genes in the F_1 from which it is derived, the mean of an F_3 family is unaffected by the linkage relations of the genes in its F_2 parent. Thus the means of the F_3 families will show the effects of linkage only by virtue of the frequencies with which the different genotypes appear in F_2, that is in exactly the same way as does the F_2 itself. Hence V_{1F2}, V_{1F3} and by derivation W_{1F23} all depend on the same D and H, which themselves reflect the recombination that occurred at gametogenesis in the F_1. When we turn to the mean variance of the F_3 families the situation is different. The frequencies of the different types of F_3 family each with its own variance, will of course reflect recombination at gametogenesis in the F_1, but their individual variances, at least in the families derived from doubly heterozygous F_2 individuals, will reflect recombination at gametogenesis in the F_2. Thus the mean variance of F_3 is of rank 2, because it shows the effects of two rounds of recombination, just as rank 1 variances showed the effects of only one round of recombination. It is not surprising therefore that while the mean variance of F_3 can still be written as $V_{2F3} = \frac{1}{4}D + \frac{1}{8}H + E$ the definition of D and H have changed to

$$D = d_a^2 + d_b^2 \pm 2(1-2p)^2 d_a d_b$$

and $$H = h_a^2 + h_b^2 + 2(1-2p)^2 (1-2p+2p^2) h_a h_b.$$

The same definition will apply to V_{2F4} and W_{2F34}, the rank 2 statistics of F_4, just as the rank 1 definition will apply to V_{1F4} and W_{1F34}. The mean variance of F_4 families will, however, by extension of the argument reflect three rounds of recombination, at gametogenesis in F_1, F_2 and F_3, and hence will be of rank 3 as is denoted by it being written as V_{3F4}. The rank 3 components of variation which appear in V_{3F4} are

$$D = d_a^2 + d_b^2 \pm 2(1-2p)^3 d_a d_b \text{ and}$$

$$H = h_a^2 + h_b^2 + 2(1-2p)^2 (1-2p+2p^2)^2 h_a h_b.$$

When we turn to the back-crosses we find that, as might now be expected, while $V_{B1} + V_{B2}$ can still be written as $\frac{1}{2}D + \frac{1}{2}H + E$ the definitions of D and H reflect the effects of the linkage, being $D = d_a^2 + d_b^2 \pm 2(1-2p)\, d_a d_b$ and $H = h_a^2 + h_b^2 + 2(1-2p)\, h_a h_b$. The definition of D is the same as in V_{1F2} but that of H is different from any that we have seen before. If we go on further to the generations derived from the back-crosses we find the same thing: the effects of linkage are accommodated by characteristic changes in the definitions of D and H which reflect the number of rounds of recombination, just as they do in the generations derived from F_2.

Unlike non-allelic interaction, linkage cannot of itself be detected and its effects measured by the analysis of means. We must go directly to second degree statistics for this purpose. Before we can do so, however, we must generalize the results we have obtained from the combination of two genes to cover any number of them. Now for every two genes there will be a potential term in $d_a d_b$ and another in $h_a h_b$, that in $d_a d_b$ taking sign according to the phase of linkage, coupling or repulsion. We can thus write general expressions for D and H in the form

$$D = S(d_a^2) + S[\pm 2(1-2p) d_a d_b] \quad \text{and}$$

$$H = S(h_a^2) + S[2(1-2p)^2 h_a h_b]$$

for rank 1 variances and covariances; in the form

$$D = S(d_a^2) + S[\pm 2(1-2p)^2 d_a d_b] \quad \text{and}$$
$$H = S(h_a^2) + S[2(1-2p)^2(1-2p+2p^2)h_a h_b]$$

for rank 2, and so on for rank 3 components and for back-crosses and so on. With k genes there will be k items each to sum in $S(d^2)$ and $S(h^2)$ and $\frac{1}{2}k(k-1)$ items in $S[\pm 2(1-2p)\, d_a d_b]$ etc. and $S[2(1-2p)^2 h_a h_b]$ etc.

The linkage of a number of genes will exert its maximum effect when all are coupled and all their h's are reinforcing. All the terms in p will then be positive. The consequences of repulsion and opposition can never be so great, except in the special case of two genes, since more than two genes can be neither all repulsed nor have their h's all in opposition. The maximum effects of repulsion and opposition might be expected when all are linked, the adjacent genes along the chromosome being repulsed and their h's opposed. Even then the 1st, 3rd, 5th, etc. will be coupled and reinforcing, as must the 2nd, 4th, 6th, etc. Inequality of the d's and h's of the various genes will also reduce the effect of linkage on the components of variation.

Even though linkage was in fact present, its effect on the value of a statistic could be zero, since the coupling and repulsion items could balance out as also could reinforcement and opposition. The balance will obviously depend on the magnitudes of effect of the genes and on the recombination frequencies, in addition to the phasic relations of the linkage. Furthermore, even where a balance is struck in the first rank components of variation the items in the components of other ranks will not balance so exactly. The effect of linkage may still thus appear, although in such a case it must be expected to be very small.

The test for linkage is thus basically a test of homogeneity of the D and H components of variation over rank. In the absence of linkage these components should be as homogenous between ranks as within them. With linkage operating, the components should be heterogenous between ranks by comparison with their variation within ranks. This test is seen at its simplest by reference to a study of ear-conformation in barley, described by Mather (1949). Ear-conformation was measured by an index compounded of ear-length, ear width and the density of the spikelets in the centre of the ear. A cross was made between two varieties, Spratt and Goldthorpe, each of which was effectively true-breeding (as indeed varieties of barley normally are) from which an F_2 and F_3 were raised. The parents thus provide estimates of the non-heritable variation E_1, between individuals within the plots of ten plants into which the experiment was divided, and E_2 the non-heritable variation between the means of plots. Each of the 100 F_3 families occupied one plot. V_{1F2}, W_{1F23}, V_{1F3} and V_{2F3} were calculated, V_{1F2} being found as the variance of F_2 individuals within the ten plots allocated to it in each of the five blocks into which the experiments was divided. The values of V_{1F2}, W_{1F23}, V_{1F3}, V_{2F3}, E_1 and E_2 averaged over the five blocks are shown in Table 38, together with their expectations in terms of the components of variation.

TABLE 38.

Ear conformation in barley (Mather, 1949). D_1 and H_1 denote the rank 1 components, and D_2 and H_2 the rank 2 components. E_1 and E_2 are the non-heritable variances of individuals and family means respectively

Statistic	Observed	Expectation	Heritable variation		
			Observed	Expected	Deviation
V_{1F2}	9713	$\frac{1}{2}D_1+\frac{1}{4}H_1+E_1$	8492	8489	3
W_{1F23}	6833	$\frac{1}{2}D_1+\frac{1}{8}H_1$	6833	6844	−11
V_{1F3}	6247	$\frac{1}{2}D_1+\frac{1}{16}H_1+E_2$	6028	6021	7
V_{2F3}	4313	$\frac{1}{4}D_2+\frac{1}{8}H_2+E_1$	3093	4244	−1151
E_1	1221	——			
E_2	219	——	$D_1 = 103\ 97$	$D_1+\frac{1}{2}H_1 = 169\ 77$	
			$H_1 = 131\ 60$	$D_2+\frac{1}{2}H_2 = 123\ 72$	

A proper analysis of these results requires the use of least squares techniques, an unweighted form of which was used by Mather (loc. cit.). A much simpler analysis will, however, serve to bring out the points in which we are interested. We can first correct for the non-heritable variation by subtracting E_1 from V_{1F2} and V_{2F3}, and E_2 from V_{1F3}. The results of doing so are shown in the fourth column of the table. Thus corrected, V_{1F2} supplies an estimate of $\frac{1}{2}D_1 + \frac{1}{4}H_1$, V_{1F3} an estimate of $\frac{1}{2}D_1 + \frac{1}{16}H_1$ and W_{1F23} an estimate of $\frac{1}{2}D_1 + \frac{1}{8}H_1$ where D_1 and H_1 denote the first rank forms of D and H. We can thus find $W_{1F23} + 2V_{1F3} - V_{1F2} = D_1 =$ 103 97. Then $V_{1F2} + V_{1F3} + W_{1F23} - \frac{3}{2}D_1 = \frac{7}{16}H_1 = 5757.5$ giving $H_1 =$ 131 60.

These joint estimates of D_1 and H_1 allow us to formulate expectations for the heritable portions of V_{1F2}, W_{1F23} and V_{1F3} as set out in the fifth column of the table, and the agreement between expectation and observed values is very close. On the assumption that there is no linkage and that the second rank components, D_2 and H_2, will be the same as those of the first rank, D_1 and H_1, we can also use these same estimates to formulate an expectation for V_{2F3}. This expectation, also shown in the fifth column of the table is 4244, while the value actually observed was 3093, a difference of 1151. Thus while agreement within the rank 1 statistics is good, agreement for the rank 2 statistic is very poor. Evidently D and H are homogeneous over rank 1 statistics but heterogeneous between ranks 1 and 2. Linkage must be operating.

The analysis can be taken a step further. Reverting for a moment to the case of two genes, A-a and B-b, with coupling $D_1 = d_a^2 + d_b^2 + 2(1-2p)\, d_a d_b$ and $D_2 = d_a^2 + d_b^2 + 2(1-2p)^2\, d_a d_b$. Both are greater in value than $D = d_a^2 + d_b^2$ which obtains in the absence of linkage. Also $D_1 > D_2$ since $(1-2p) > (1-2p)^2$. Thus with coupling the value of D will fall from rank 1 statistics to rank 2. With repulsion there would be a corresponding rise. Furthermore, expressed as a proportion of D_1 the fall will be

$$\frac{D_1 - D_2}{D_1} = \frac{[d_a^2 + d_b^2 + 2(1-2p)\, d_a d_b] - [d_a^2 + d_b^2 + 2(1-2p)^2\, d_a d_b]}{d_a^2 + d_b^2 + 2(1-2p)\, d_a d_b}$$

which reduces to

$$\frac{4p(1-2p)}{4(1-p)} \text{ when } d_a = d_b.$$

This ratio of the fall to D_1 is at its maximum value of 0.17 when $p =$ 0.29.

We cannot however compare D_1 and D_2 from the barley experiment, because with only one rank 2 variance we cannot separate D_2 and H_2. We must therefore work in terms of $D + \frac{1}{2}H$ upon which the heritable components of V_{2F3} depend. V_{2F3} yields us a joint estimate $D_2 + \frac{1}{2}H =$ $4 \times 3093 = 123\,72$. The first rank statistics yield $D_1 + \frac{1}{2}H_1 = 103\,97$ $+ \frac{1}{2}(131\,60) = 169\,77$ which is markedly larger than $D_2 + \frac{1}{2}H_2$. There must therefore be linkage in coupling, with a fall ratio

$$\frac{(D_1 + \frac{1}{2}H_1) - (D_2 + \frac{1}{2}H_2)}{D_1 + \frac{1}{2}H_1} = \frac{460\,5}{169\,77} = 0.27.$$

Now with the h's reinforcing H_1 will be greater than H_2, by $2\, h_a h_b$ $(1-2p)^2\, [1-(1-2p+2p^2)]$. If $h_a = h_b$ this fall is 0.06 of H_2 when $p = 0.29$. If we assume that $h_a = h_b = d_a = d_b$, as is statistically consistent with the data, the fall ratio of $D + \frac{1}{2}H$ with 2 genes at $p = 0.29$ is 0.13. Although the maximum fall ratio in H is at a somewhat lower value of p than is that in D, this value 0.13 is a sufficiently good approx mation to the maximum for our purpose, because it is only about half the fall ratio actually observed. Clearly two coupled genes are incapable of explaining the barley results. As the number k of linked genes increases, however, the number of terms in $d_a d_b$ will increase as $\frac{1}{2}k(k-1)$ They will therefore loom larger in the composition of D, and the fall ratio will increase correspondingly, where all the genes are coupled. It is possible to calculate the maximum fall-ratio given by three or more genes just as we did for two, and when this is done we find that a minimum of about four coupled genes is required to give a fall ratio of 0.27 as found in the barley. In point of fact, for reasons into which we need not enter here, the polygenic system governing ear-conformation in this barley cross is likely to be more complex even than this simple consideration of the fall ratio would indicate (Mather, 1949).

Although scaling tests applied to the means of parents, F_1 F_2 and F_3 in the barley revealed some evidence of non-allelic interaction (probably arising from inadequacy of the scale), this was clearly too small to account for the heterogeneity of D and H, a conclusion which is further substantiated by the homogeneity of V_{1F2}, W_{1F23} and V_{1F3}. Where, however, the scaling tests have revealed major interaction, the test of linkage becomes more difficult. Interaction produces heterogeneity of D and H over generations, but not within them: linkage gives heterogeneity of D and H between ranks, but not within them. Difficulty arises, however, because generation and rank are themselves related, since an additional rank can be obtained only by introducing an additional generation. In principle an unambiguous test is possible if a sufficiently complex crossing programme is used (Van der Veen, 1959), and Perkins and Jinks (1970) have been successful in obtaining conclusive evidence of linkage in the presence of interaction using generations of less familiar types. The whole subject is however, complex and worthy of more study than it has yet received.

23. Diallels

The means of the families which constitute a set of diallel crosses will reflect any interaction shown by the genes in which the parental lines differ. On the other hand, since only these means are used in diallel analysis, and indeed the families themselves are non-segregating in the diallels we have been observing, linkage as such can be having no effect on the variation that we observe and measure. At the same time the genes in which the parental lines differ may be correlated in their distributions among the parents and in such a case their contributions to the variation among the families of the diallel will not be independent.

The general expression for the effects of digenic interaction on the means, variances and covariances of a diallel are very complex (M and J, Table 96). We can, however, learn something of the ways in which both interaction and correlated gene distributions express themselves in diallel analysis if we consider the special and relatively simple case of four parental lines representing all the combinations of two genes pairs with $u_a = v_a = u_b = v_b = \frac{1}{2}$ (i.e. all gene frequencies equal) but having correlated distributions among the four parents, and where $d_a = d_b = h_a = h_b$ and $i = j_a = j_b = l = \theta d$ (i.e. with digenic interaction of the complementary-duplicate type). The correlation of the gene distributions is measured by the parameter c the frequencies of the AABB and aabb parents each being $\frac{1}{4}(1 + c)$ and those of the AAbb and aabb parents each being $\frac{1}{4}(1 - c)$. When $c = 0$, all the parents occur with the frequency $\frac{1}{4}$. When $c = 1$ association is complete, the AAbb and aaBB parents being absent, with A and B on the one hand and a and b on the other always occurring together as a single compound gene pair. Equally when $c = -1$ dispersion is complete, A always occurring with b and a with B the AABB and aabb parents being absent. Values of c between 1 and -1 represent various strengths of association and dispersion.

Similarly the interaction is measured by θ. So with $d_a = d_b$ and $i = \theta d$ the phenotype of for example AABB, which in general terms is $d_a + d_b + i$, can be written as $d(2 + \theta)$, and that of AAbb as $d(-\theta)$. Similarly with $h = d$ and $l = i = \theta d$ the phenotype of AaBb, which in general terms is $h_a + h_b + l$, becomes $d(2 + \theta)$ and so on. The phenotypes of the sixteen families in the diallel are set out in these terms in the body of Table 39, where the frequencies of the four parental lines are also shown in terms of c.

The diallel table is sufficiently simple for us to undertake a full analysis. The first point to note is that since $d_a = d_b = h_a = h_b$ and $j_a = j_b$, the central two arrays will be alike in the values they yield for V_r and W_r and so will provide only a single joint point in the W_r/V_r graph, which thus will have only three points instead of the more general four. The mean of the parents will be $\frac{1}{4}d[(1 + c)(2 + \theta) - 2(1 - c)\theta + (1 + c)(-2 + \theta)] = d\theta c$ and the mean of array ab will obviously be the same. Since the phenotype is $d(2 + \theta)$ for all four classes in the AB array, its mean will obviously be $d(2 + \theta)$ while the means of the Ab and aB arrays will be $\frac{1}{4}d[(1 + c)(2 + \theta) - (1 - c)\theta + (1 - c)(2 + \theta) - (1 + c)\theta] = d$. V_r for

TABLE 39.

Two-gene diallel set of matings with complementary/duplicate interaction, measured by θ, and equal gene frequencies but correlated gene distributions, measured by c. The body of the table gives the phenotypes of the various classes in terms of θ and the frequencies of the parents are shown in terms of c. $d_a = d_b = h_a = h_b = d$

	Frequency	Genotype and phenotype	Female parent AABB	AAbb	aaBB	aabb
Male parent	$\frac{1}{4}(1+c)$	AABB $d(2+\theta)$	AABB $d(2+\theta)$	AABb $d(2+\theta)$	AaBB $d(2+\theta)$	AaBb $d(2+\theta)$
	$\frac{1}{4}(1-c)$	AAbb $d(-\theta)$	AABb $d(2+\theta)$	AAbb $d(-\theta)$	AaBb $d(2+\theta)$	Aabb $d(-\theta)$
	$\frac{1}{4}(1-c)$	aaBB $d(-\theta)$	AaBB $d(2+\theta)$	AaBb $d(2+\theta)$	aaBB $d(-\theta)$	aaBb $d(-\theta)$
	$\frac{1}{4}(1+c)$	aabb $d(-2+\theta)$	AaBb $d(2+\theta)$	Aabb $d(-\theta)$	aaBb $d(-\theta)$	aabb $d(-2+\theta)$
		Array mean	$d(2+\theta)$	d	d	$d(\theta c)$
		V_r	0	$d^2(1+\theta)^2$	$d^2(1+\theta)^2$	$d^2(2+\theta^2+2c)$
		W_r	0	$d^2(1+\theta)(1+c)$	$d^2(1+\theta)(1+c)$	$d^2(2+\theta^2+2c)$
		W_r+V_r	0	$d^2(1+\theta)(2+\theta+c)$	$d^2(1+\theta)(2+\theta+c)$	$2d^2(2+\theta^2+2c)$
		W_r-V_r	0	$d^2(1+\theta)(c-\theta)$	$d^2(1+\theta)(c-\theta)$	0

the AB array will clearly be 0 since the phenotypes of all its classes will be alike, and so of course will its W_r also. For the ab array

$$V_r = \tfrac{1}{4}d^2[(1+c)(2+\theta)^2 + 2(1-c)\theta^2 + (1+c)(-2+\theta)^2] - d^2\theta^2c^2 = d(2+\theta^2+2c)$$

the term $d^2\theta^2c^2$ being the correction for the mean. The variance of the parents will obviously be the same as V_r for the ab array, since the phenotypes of the four classes in the array are the same as those of the corresponding parents. For these reasons also W_r will equal V_r for this array. Turning to the central arrays we find

$$V_r = \tfrac{1}{4}d^2[(1+c)(2+\theta)^2 + (1-c)(-\theta)^2 + (1-c)(2+\theta)^2 + (1+c)(-\theta)^2] - d^2 = d(1+\theta)^2$$

the d^2 being the correction for the mean. Similarly for these two arrays

$$W_r = \tfrac{1}{4}d^2[(1+c)(2+\theta)^2 + (1-c)(2+\theta)(-\theta) + (1-c)(-\theta)^2 + (1+c)(-2+\theta)(-\theta)] - d^2\theta c = d^2(1+\theta)(1+c).$$

These various results are collected together in the lower part of Table 39, as are $W_r + V_r$ and $W_r - V_r$ for each array.

A number of conclusions emerge from these results. In the first place $W_r - V_r = 0$ for both the AB and ab arrays. Their points in the W_r/V_r graph will thus be on a line of slope 1 which passes through the origin, no matter what the situation may be about the interaction and gene distribution. Furthermore since this line intercepts the ordinate at the origin it indicates that $d = h$, which of course agrees with the assumption on which the analysis is based. The point from the central arrays Ab and aB, will however lie on this line only when $W_r - V_r = d^2(1+\theta)(c-\theta) = 0$ and this will happen only when $c = \theta$. When both interaction and correlation of gene distribution are absent, $c = \theta = 0$ and a straight line is obtained for the regression of W_r and V_r, after due correction has been made for any non-heritable variation, as indeed we saw in Chapter 4. This will also happen in the presence of both interaction and correlated distribution provided that, as measured by θ and c respectively, they are equally strong.

Either the interaction by itself ($\theta \neq 0$, $c = 0$) or correlation of the gene distribution by itself ($\theta = 0$, $c \neq 0$) must result in the regression of W_r on V_r departing from a straight line of slope 1. The relation of this departure to the strength of the interaction (θ) is shown in Fig. 13 and to the strength of the correlation (c) in Fig. 14. The values of the W_r and V_r are divided by $d^2(2+\theta^2)$ in the one case and $d^2(2+2c)$ in the other in order to standardize the graph by making the point for the ab array fall at $W_r = V_r = 1$. When θ is positive (complementary type interaction) or c is negative (dispersion of the genes) the central point lies to the right of and below the line of slope 1 through the origin delimited by the points from array AB and ab. When θ is negative (duplicate type interaction) or c is positive (association of the genes) it lies above and to the left of the line. The relation of the departure from the line to the value of θ or c is shown by the trajectory the central point follows with change in θ or c. These trajectories are not the same for interaction and correlated distribution of the genes. Since however the interactive properties of two genes are presumably fixed and their gene distributions are equally fixed in any set of parents, we can obtain only one point in the trajectory and so the difference in trajectories is of no help to us in seeking to distinguish the effects of interaction from those of association

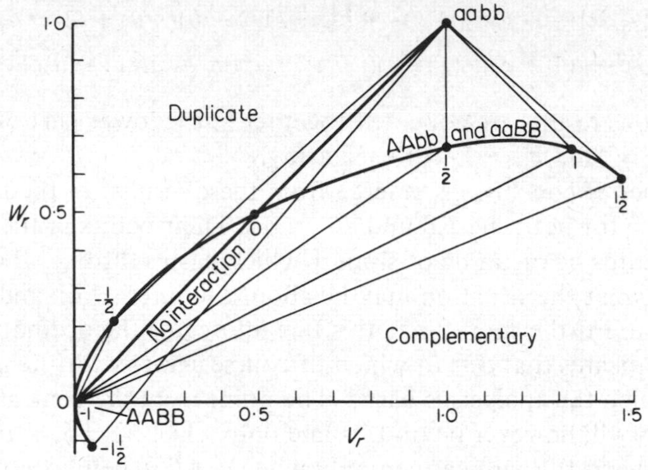

Fig. 13. The effect of complementary and duplicate type interaction, measured by θ, between two gene pairs on the W_r/V_r graph from a diallel set of matings, with $d_a = d_b = h_a = h_b$ and $c = 0$. The AAbb and aaBB arrays give a common point which lies mid-way on a straight line between the AABB and aabb points. Complementary interaction causes this point to move to the right and upwards and the W_r/V_r graph ceases to be a straight line, becoming concave upwards. Duplicate interaction produces the opposite result, the graph becoming concave downwards. The heavy curve shows the path of the middle point as it moves under the influence of interaction, the numbers indicating the values of θ to which the points correspond.

or dispersion. In short, diallel analysis enables us to detect interaction and/or correlation of the gene distributions but it does not enable us to distinguish between them.

When both interaction and correlation of the gene distribution are present, they may either reinforce one another's action in moving the central point away from the line if θ and c are of opposite sign, or oppose one another's effects if θ and c are of the same sign. They will balance exactly and the central point will fall on the line itself whenever $\theta = c$.

We thus see how interaction and correlation of the gene distributions can affect the W_r/V_r graph in ways which are not distinguishable on the basis of this evidence alone, and how they can reinforce, oppose and even cancel out one another's effects on the graph. In conclusion it should be remembered that we have been considering only the special case of complementary-duplicate type interaction with equal gene frequencies. Our findings still hold good when the gene frequencies are

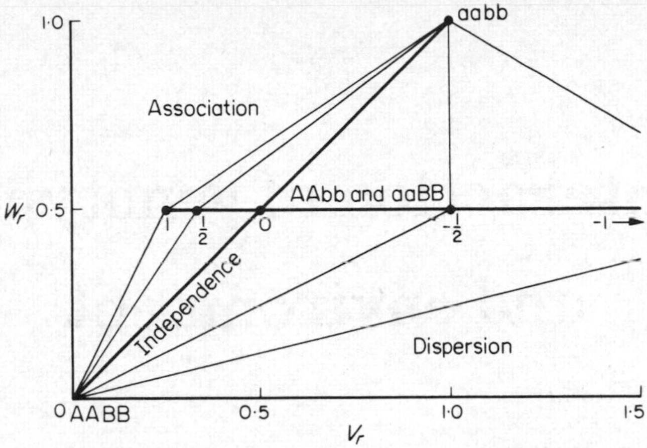

Fig. 14. The effect of gene association and dispersion, measured by c, of two gene pairs on the W_r/V_r graph from a diallel set of matings, with $d_a = d_b = h_a = h_b$ and $\theta = 0$. The effect of association is similar to that of duplicate interaction and dispersion to that of complimentary interaction, illustrated on Fig. 13. The path of the middle point with change in c is not however curved, as with interaction, but follows a line parallel to the abscissa as shown by the heavy line. The numbers indicate the values of c to which the points correspond.

unequal. Indeed they apply generally in respect of correlated gene distributions, and they are unlikely to be modified in more than detail in respect of this general type of interaction where the d's and h's are not equal, although the graph will then have four points on it, not just three as in the special case we have been discussing. We should not, however, extrapolate these to other less simple systems of interaction, which although yet to be fully investigated are known to be capable of producing very bizarre effects on the W_r/V_r graph.

6

Interaction of genotype and environment

24. Genotype × environment interaction

The simple additive-dominance model assumes that gene differences contribute independently from one another to variation in the phenotype. We have seen how failure of this assumption can be detected and how departures from the model may be produced by the interaction of non-allelic genes and by the correlation of gene distributions, both of which may be described in terms of appropriate parameters whose values can be estimated from suitable data. As we have developed and used it so far, the additive-dominance model further assumes that gene differences and environmental differences also contribute independently of one another to variation in the phenotype. We must now turn to consider the interaction of gene and environmental differences (or genotype × environment interaction as it is commonly called), how such interaction may arise, and how it can be detected, measured and investigated.

Genotype × environment interaction has long been known to occur. An early example is that of Åkerman (1922), who reported a genetic difference affecting the chlorophyll of oats which was undetectable when the plants were grown in subdued light but revealed itself by the bleaching and death of one genetic class when they were grown in direct sunlight. Interaction of genotype and environment must indeed be expected to occur and in fact some gene changes must themselves result in marked changes of the environment which the individuals experience; climbing beans and dwarf beans, for example, must experience very different environments although they may differ in only a single gene. This is, however, an extreme example and we must expect most cases of genotype × environment interaction to be much less dramatic. We must therefore seek to give a more general account of them and to develop an appropriately general method for their investigation.

Like other forms of interaction, that between genotype and environment may arise from the scale on which the character is measured and represented. An example of this is afforded by the data from Hogben (1933) quoted by Mather and Jinks (1971) concerning the average numbers of facets in the eyes of two strains of *Drosophila melanogaster*, referred to as Low-Bar (L) and Ultra-Bar (U) raised at two temperatures, 15 and 25° C. These facet numbers are shown diagrammatically on the left of Fig. 15. At 15° C, L had on average 146 more facets than U, but

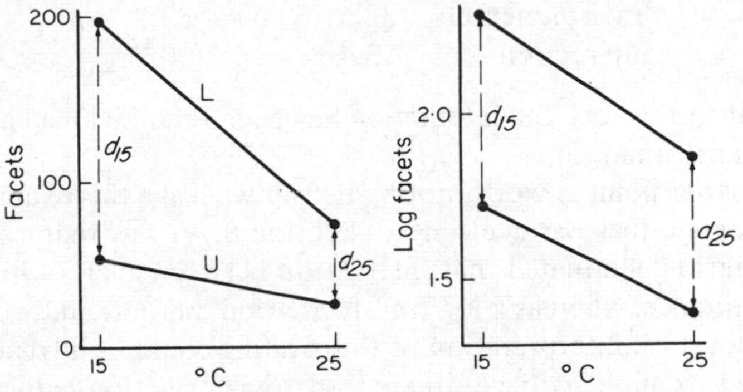

Fig. 15. Krafka's data (from Hogben, 1933) on the mean numbers of facets in the eyes of two lines of Bar-eyed *Drosophila* at two temperatures. When the direct count of eye facets is used (on the left) the difference between the lines at 15° C (d_{15}) is larger than the difference at 25° C (d_{25}), so indicating genotype × environment interaction. When, however, the logs of the mean numbers of eye facets are used (on the right) d_{15} and d_{25} are nearly equal. The scalar transformation has removed the interaction.

at 25° C the difference is only 49. At the higher temperature the difference is only 1/3 of that at the lower. The lines are not reacting equally to the change in temperature: the effects of genotype and environment are not additive, or in other words, there must be an interaction of genotype and environment. When, however, we change the scale by taking logarithms of the number of facets, we obtain the picture shown on the right of Fig. 15. In log measure the difference between L and U is 0.58 at 15° C and 0.47 at 25° C. The higher temperature still gives a smaller difference than the lower, but the reduction is proportionately very much less than when the untransformed facet number was used. The log transformation has very much reduced the genotype × environment interaction, if not entirely eliminated it.

The size of the reduction emerges even more dramatically if we subject the data to an analysis of variance. The 3 df among the four observations may be assigned 1 each to the overall effect of the genetic difference, the overall effect of the environment, and the genotype × environment interaction. The percentages of the total variation taken out by each of these three items using direct measure and log measure are:

Item	*Direct-Measure-Log*	
Genetic	54.1	66.1
Environmental	32.5	33.2
Interaction	13.4	0.7

Looked at in this way the interaction has been rendered negligible by the log transformation.

One further point is worth noting before we leave this example. When considering another Bar-eye gene, in Section 8, we saw that a square root transformation eliminated that interaction between alleles which we term dominance, whereas a log transformation did not, and we saw too that a theoretical interpretation of this finding could be advanced. In the present example, while a square root transformation reduces the genotype × environment interaction it is much less effective than the log transformation. This contrast emphasizes the essentially empirical nature of choice of a transformation, and the unwisdom of seeking to draw theoretical conclusions from a successful case of a particular change of scale.

Not all genotype × environment interactions can, however, be ascribed to the use of an inappropriate scale for the representation of the character. Table 40 sets out the mean numbers of sternopleural chaetae borne by the two inbred lines, Samarkand (S) and Wellington (W), of *Drosophila* when raised in six different environments, which comprised all the possible combinations of two temperatures 18 and 25° C, and three types of culture vessel, $\frac{1}{3}$ pint milk bottles with yeasted food (B), 1 × 3 inch vials with yeasted food (Y), and similar vials with unyeasted food (U). Five cultures were reared of each line in each environment, the figures in the table being the means of all the five replicate cultures in each case. Comparisons among the five replicates give us an estimate of error variation which will be based on 4 df within each combination of genotype and environment. Since there are 2 × 6 = 12 such combi-

TABLE 40.

Mean numbers of sternopleural chaetae in the S and W inbred lines of *Drosophila melanogaster*, their F_1 and F_2 raised in six environments

	18°C			25°C			Error variance
	B	Y	U	B	Y	U	
S	20.58	20.51	20.26	20.44	20.93	20.66	0.020 721
W	19.63	19.34	19.34	18.67	18.14	17.61	0.020 721
F_1	19.98	20.01	20.16	19.22	18.93	18.48	0.010 332
F_2	20.19	19.86	19.75	19.45	18.68	18.75	0.101 823

The environments are the six possible combinations of two temperatures, 18 and 25°C, with three types of culture, in 1/3 pint milk bottles with yeasted food (B), and in 3 X 1 inch vials with yeasted (Y) and unyeasted (U) food.

nations, the pooled estimate of error variation will thus be based on 48 df. It turns out to be 0.1036. As the entries in the table are the means of five replicates, they will be subject to an error variance of $0.1036 \div 5 = 0.020\,72$, as shown in the right-hand column.

We might note in passing that the mean numbers of chaetae were also determined for the F_1 and F_2 of the cross between these two lines. Although they will not be discussed until later, these means are also recorded in Table 40. Equal numbers of families were raised from the reciprocal crosses, S X W and W X S in both F_1 and F_2. Eight replicates, four from each reciprocal, were raised of the F_1 in each environment, but only two, one from each reciprocal, of the F_2. The entries for F_1 are thus the means of eight replicates and those for F_2 the means of two. The error variances of their entries were found separately for F_1 and F_2 although otherwise in the same way as for S and W themselves, and are given in the table. Not surprisingly the error variance of the F_1 entries is lower than that of the parents, but that of the F_2 entries is much higher.

Returning to the parent lines, the numbers of chaetae of S averaged over B, Y and U is $\frac{1}{3}(20.58 + 25.51 + 20.26) = 20.45$ to two places of decimals, at 18°C and $\frac{1}{3}(20.44 + 20.93 + 20.66) = 20.68$ at 25°C. Those for W are similarly 19.44 and 18.14. Thus W's chaeta number is 1.30 higher at 18 than at 25°C while that of S changes much less, such change as there is being a reduction of 0.23, i.e. in the opposite direction to W. Clearly the lines are reacting differently to the change in temperature. Since, however, the change in W is a major reduction with increase in temperature, while that in S is if anything in the opposite direction, no simple or even acceptable transformation of the scale on which chaeta number is measured could eliminate this apparent interaction of the two genotypes with the environmental difference. Clearly given that it is significant we must accept the interaction as it is and elaborate our model to accommodate it.

25. Two genotypes and two environments

Now if we let $[d]$ be the genetically determined deviation of $\bar{S}$, the mean chaetae number of S from the mid-parent, m, and $-[d]$ that of $\bar{W}$, the assumption made by the simple model, that the non-heritable deviations, spring from the environmental difference are independent of the genotype, would be tantamount to saying that the environment adds a deviation e at 18°C and a deviation $-e$ at 25°C, equally in the cases of both genotypes. The situation would then be as shown in Table 41, which sets out the algebraic formulation for the two genotypes in the two environments, with the corresponding mean number of chaetae, (rounded off to two decimal places) below them. We can proceed to estimate the parameters we have used. m is the overall average of the observations and is found as $\frac{1}{4}(20.45 + 20.68 + 19.44 + 18.14) = 19.6775$. The gen-

TABLE 41.

Mean chaeta numbers of the S and W inbred lines at 18 and 25°C

		18°C	25°C	Sum
		$m+[d]+e$	$m+[d]-e$	$2m+2[d]$
	Obs	20.45	20.68	41.13
S	Exp	20.8325	20.2975	41.13
	O-E	−0.3825	0.3825	0
		$m-[d]+e$	$m-[d]-e$	$2m-2[d]$
	Obs	19.44	18.14	37.58
W	Exp	19.0575	18.5225	37.58
	O-E	0.3825	−0.3825	0
	Sum	$2m+2e$	$2m-2e$	$4m$
		39.89	38.82	78.71
		$m = 19.6775$	$[d] = 0.8875$	$e = 0.2675$

etic parameter is estimated from the line sums in the right-hand column of the table as $[d] = \frac{1}{4}(41.13 - 37.58) = 0.8875$, and the environmental parameter is similarly found from the environmental sums in the bottom row of the table as $e = \frac{1}{4}(39.89 - 38.82) = 0.2675$. We can now construct expected values for the chaeta numbers of the two lines at the two temperatures by substituting the estimates of m, $[d]$ and e in the formulations that the model yields. Thus the expected chaeta number (E) of W at 25° C is $m - [d] - e = 19.6775 - 0.8875 - 0.2675 = 18.5225$ which compares with 18.14, the number observed (O), giving a difference $O - E = -0.3825$. When comparing the expectation so obtained with the observed chaeta numbers we find that S at 18° C also gives $O - E = -0.3825$, while S at 25 and W at 18° C give a difference $O - E = 0.3825$. The large size of these deviations relative to $[d]$ and e suggests strongly that the simple model we have used is inadequate and that the two genotypes do not react equally to the change in temperature: in other words that genotype X environment interaction is present. We can accommodate this interaction by introducing a further parameter, g, into the formulation in the way shown in the upper expressions of Table 42. This new parameter g is a measure of the genotype by environment interaction and in the present case is estimated as $g = \frac{1}{4}(20.45 - 19.44 - 20.68 + 18.14) = -0.3825$. In conjunction with $[d]$ and e, g completes the set of three parameters, corresponding

TABLE 42.

Alternative models for the phenotypes given by two genotypes, S and W, raised in two environments, 18 and 25°C

	18°C	25°C
S	$m + [d] + e + g$ $m + [d] + e_s$	$m + [d] - e - g$ $m + [d] - e_s$
W	$m - [d] + e - g$ $m - [d] + e_w$	$m - [d] - e + g$ $m - [d] - e_w$

In each case the upper expression is in terms of the genetical parameter $[d]$ found by averaging over environments, the environmental parameter e found by averaging over genotypes, and g the statistical interaction of $[d]$ and e. The lower expression is in terms of the same genetical parameter $[d]$, but with e_s measuring the change in expression of genotype S between the environments and e_w similarly measuring the change in expression of W. Three parameters are involved in each formulation, $[d]$ being the same in both, with $e_s = e + g$ and $e_w = e - g$.

to the 3 df among four observations, required to give a perfect fit for the deviation of the four observed chaeta numbers from their mean, m.

In the absence of interaction g will not depart significantly from 0, and we can therefore test the adequacy of the simple model, which assumes no interaction, by testing the significance of g. This can be done in either of two ways, which both give the same answer. First, since $g = \frac{1}{4}(\bar{S}_{18} - \bar{W}_{18} - \bar{S}_{25} + \bar{W}_{25})$ where $\bar{S}_{18}$ is the mean chaeta number of S at 18° C, etc.

$$V_g = \tfrac{1}{16}(V_{\bar{S}18} + V_{\bar{W}18} + V_{\bar{S}25} + V_{\bar{W}25}) \quad \text{and} \quad s_g = \sqrt{V_g}.$$

Each chaeta number in Table 42, from which g has been calculated, is the mean of three of the observations in Table 41, and each of these observations is subject to an error variance of 0.020 72, based on 48 df as we have already seen. Thus each chaeta number in Table 41 has an error variance of $\frac{1}{3}(0.020\,72) = 0.006\,907$ and V_g will thus be $\frac{1}{16}(4 \times 0.006\,907$ $= 0.001\,727$ giving $s_g = \sqrt{V_g} = 0.0414$. Then $t_{48} = g/s_g = 9.2$ giving a very small probability. The interaction is thus significant and the simple model must be judged to be inadequate. We may note that $[d]$ and e will have the same standard error as g, and when tested in the same way they also both depart very significantly from 0.

The second way of testing g, and also $[d]$ and e, is by an analysis of variance of the four chaeta numbers in Table 41. As we have seen $[d] = \frac{1}{4}(41.13 - 37.58) = \frac{1}{4}(3.55) = 0.8875$. The SS accounted for by $[d]$ will thus be $\frac{1}{4}(3.55)^2 = 3.1506$. Since this SS stems from a single parameter and hence corresponds to 1 df, the MS will be the same as the SS. Finding the SS's accounted for by e and g similarly, we obtain the analysis of variance shown in Table 43. This also includes the estimate of error variance applicable to the chaeta numbers which we found in the previous paragraph to be 0.006 907 and with which the MS's for the three parameters

TABLE 43.

Analysis of variance of the observations in Table 41

Item	df	MS	VR	P
$[d]$	1	3.1506	456.1	v.s.
e	1	0.2862	41.4	v.s.
g	1	0.5852	84.7	v.s.
Error	48	0.006 91		

v.s. = very small

must be compared to test their significance. Again all the three items are highly significant. Since each MS in the analysis stems from 1 df, the VR obtained when it is divided by the error variance is a t^2. Thus in the case of the g item, the VR is 84.7, giving $t = \sqrt{(\text{VR})} = 9.2$ as in the earlier test. The two ways of testing the significance of g are thus no more than two forms of the same test.

The significance of g shows that genotype × environment interaction is present, or in other words that the two genotypes S and W do not react equally to the change in temperature. This suggests an alternative formulation for the phenotypes of the two lines at the two temperatures, in which e and g are replaced by two different parameters e_s and e_w measuring respectively the differences produced in S and W by the alteration of temperature. Thus S at 18° C has the phenotype $m + [d] + e_s$ and at 25° C is $m + [d] - e_s$, while for W at the two temperatures are $m - [d] + e_w$ and $m - [d] - e_w$ as set out in the lower expressions of Table 42. This formulation has the advantage that e_s and e_w are properties of the individual lines, unlike e and g, which are compounds of the properties of the two lines. As such e_s and e_w are biologically more directly meaningful than e and g, and indeed are direct measures of the sensitivity of the two lines to change in an aspect of the environment. They thus measure a character which is prospectively important and whose genetic basis can be investigated in a direct way.

Now $[d]$, e_w and e_s permit a complete specification of the phenotype as do $[d]$, e and g. Clearly therefore, since $[d]$ is common to both formulations, e_s and e_w must relate to e and g. In fact, $e_s = e + g$, and $e_w = e - g$, or put the other way round $e = \frac{1}{2}(e_s + e_w)$ while $g = \frac{1}{2}(e_s - e_w)$, and the SS jointly accounted for by e_s and e_w equals that jointly accounted for by e and g, each SS corresponding of course to 2 df. Thus given the values of one pair of parameters the values of the other two can be found: they are no more than alternative ways of representing the same thing and are readily converted into each other. The formulation to be used may be chosen by its convenience for the investigation or analysis in hand. In general, while e_s and e_w are the more biologically meaningful pair, e and g are commonly the more analytically useful, although this is not always the case.

In the present example $e_s = \frac{1}{2}(20.45 - 20.68) = e + g = 0.2675 - 0.3825 = -0.115$ while $e_w = \frac{1}{2}(19.44 - 18.14) = e - g = 0.2675 - (-0.3825) = 0.650$. We note that e_s and e_w are each found as half the difference between two of the observed values in Table 41 each of which has an error variance of 0.006 907. Hence $V_{es} = V_{ew} = \frac{1}{4}(2 \times 0.006\,907)$

$= 0.003\,454$ and $s_{es} = s_{ew} = \sqrt{0.003\,454} = 0.058\,77$. The difference between e_s and e_w is significant (which is, of course, an alternative way of demonstrating genotype $\times$ environment interaction and leads in fact to exactly the same test of significance that we have already used), and e_w is significantly greater than 0, but e_s is not significantly negative on these results. Thus while we can say that the two genotypes respond differently to the change in temperature, we cannot say from these data that they respond in different directions.

26. A more complex case

So far we have been discussing the simplest case of two genotypes and we have derived two different approaches to the detection and measurement of the interaction. The first using $[d]$, e and g, leads to an analysis of variance into items for the effects of the genetic difference, the environmental difference and their interaction, in the familiar statistical way. The second, using $[d]$, e_s and e_w, depends on finding and comparing the changes produced by the temperature difference in the two lines taken individually. Both approaches are readily generalized to deal with any number of lines in any number of environments.

Table 40 gives the mean numbers of sternopleural chaeta not only for S and W but for their F_1 and F_2 also. Strictly we should not bring either the F_1 data nor that from F_2 into the same analysis of variance as S and W since the observations on them are subject to error variances different from that of the two parents. The error variance of the F_1 observations, however, differs from that of S and W only by a factor of two which is not likely to lead to problems of interpretation if we include them in the same analysis especially if we are conservative and assume that the parental lines error variance applies to F_1 as well. The F_2 results on the other hand have an error variance greater by a factor of five than the parents and will be excluded from the analysis of variance for this reason.

Taking S, W and F_1 we have observations on three lines in six environments, or eighteen observations which will of course yield 17 df in the analysis. Of these 2 df will correspond to the genetic difference between the three lines, and 5 df to the differences among the six environments. The remaining $2 \times 5 = 10$ will correspond to the interaction of the two main effects. The genetical items depend on differences analogous to $[d]$, in the simple case, the environmental items to differences of type e and

TABLE 44.

Analysis of variance of the observations on S, W and F_1 in Table 40

Item	df	MS	VR	P
Lines (L)	2	4.8163	232.4	v.s.
Environments (E)	5	0.6026	29.1	v.s.
Interaction (I)	10	0.2743	13.2	v.s.
Error	48	0.020 72		
L1 (S−W)	1	9.4519	456.2	v.s.
L2 ($S+W-2F_1$)	1	0.1806	8.7	v.s.
I1	5	0.4433	21.4	v.s.
I2	5	0.1052	5.1	0.001
E1 (18−25°C)	1	2.5163	121.4	v.s.
E2 (culture types)	2	0.1750	8.4	0.001
I1′	2	1.0739	51.8	v.s.
I2′	4	0.0669	3.2	0.05-0.01

the interaction to differences of type *g*. The simple analysis of variance of the eighteen observations is set out in Table 44, the error variance used being that pertaining to the observations on S and W in Table 40, as we have already noted. It is clear that all three items in the simple analysis of variance (set out in the upper part of the table) are very significant when tested against this estimate of error. There is thus evidence not only of genetical differences among S, W and F_1, and differences among the six environments but also of interaction between the genetic and environmental differences: the lines do not change equally as the environment alters, as indeed we have already seen to be so in the simpler case of S and W at the two temperatures.

We can take the analysis further. First we can compare the behaviour of S and W over all six environments. Thus in bottles (B) at 18° C we can find from Table 40, $S - W = 20.58 - 19.63 = 0.95$ and so on. The sum of the six differences is 10.65 which contributes $\frac{1}{12}(10.65)^2 = 9.4519$ for 1 df to the SS of 9.6325 for 2 df (yielding a MS of 4.8163) for lines (L) in the main analysis. The SS of the six differences round their mean is found as $\frac{1}{2}(0.95^2 + 1.17^2 \cdots + 3.05^2) - \frac{1}{12}(10.65)^2$ the divisors 2 and 12 being the number of observations that go into each difference and into the sum of the differences, respectively. This SS turns out to be 2.2167 for, of course, 5 df among the six differences giving a MS of $\frac{1}{5} \times 2.2167 = 0.4433$ for the interaction of the genetic difference between S and W with the six environments. These two items appear as L1 and I1 in the middle part of Table 44 and both are very significant, so bearing out our earlier test of the two lines over the two temperatures (Table 43), in showing that they do not react in the same way to changes of environment. The remaining comparison (L2) among the three lines is of S and W, taken together, with the F_1. We can find it as the difference between L1 and the SS for 2 df for lines in the main analysis. Thus the SS for this comparison of parents, taken together, with F_1 (L2) is $9.6325 - 9.4519 = 0.1806$ which corresponding as it does to 1 df is also the MS. The interaction item I2 is found similarly from the SS for interaction, having 10 df, in the main analysis and the SS for I1, having 5 df. The SS for I2 which also has 5 df, is thus $2.7426 - 2.2167 = 0.5259$ giving a MS of 0.1052, as entered in the middle part of the table. Although smaller than I1, I2 is also significant with a P of just about 0.001. This further analysis of the lines and interaction items shows not only that S and W differ in their overall effects on chaeta number and in their interaction with environments as well, but that the F_1 differs from the joint or mean behaviour of its parents, again in both overall effect on chaeta number and in interaction with the environments. We shall look further into these relationships in Section 28.

Just as we can compare the lines with one another over all six environments, we can compare the effects of the environments with one another over all three lines. Of the 5 df among the six environments, one can be identified as relating to the difference between the two temperatures, and two more as relating to the differences among the three types of container, B, Y and U. The effect of temperature is calculated by first finding the difference in the chaeta number at 18 and 25° C for the three lines separately. Thus the difference for W is $19.63 + 19.34 + 19.34 - 18.67 - 18.14 - 17.61 = 3.89$ those for S and F_1 being similarly -0.68 and 3.52. The overall difference between the two temperatures is hence $3.89 - 0.68 + 3.52 = 6.73$ which gives a SS of $\frac{1}{18}(6.73)^2 = 2.5163$ for 1 df, the divisor 18 being of course the number of observations of which the 6.73 is composed. This SS is part of the SS, having 5 df, for environments in the main analysis. It is shown as E1 in the lowest part of Table 44. The interaction of lines with the temperature difference (I1′) in the lowest part of Table 44, and not to be confused with I1 in the middle part of the table is found as $\frac{1}{6}[3.89^2 + (-0.68)^2 + 3.52^2] - \frac{1}{18}(6.73)^2 = 2.1479$ for 2 df giving a MS of 1.0739. This is part of the interaction SS in the main analysis of variance, and is very significant when tested against the error variance, 0.020 72.

The interaction with container types is most easily found by constructing a 3 X 3 table in which each entry is the sum of two corresponding values, one from each temperature, the nine entries being one for each container type in each line. Thus the value for B in line W is 19.63 + 18.67 = 38.30: that for U in F_1 is 20.16 + 18.48 = 30.64 and so on. An analysis of variance can then be carried out on the entries in this 3 X 3 table, an additional factor of two being used in all the divisors because each entry is the sum of two of the initial observations from Table 40. One margin of the 3 X 3 table will yield a SS for 2 df reflecting the line differences and will be exactly the same as the lines item in the main analysis of variance. The other margin yields a SS of 0.3499 for 2 df, giving a MS of 0.1750, for the overall variation between the three container environments (E2 in the lowest part of Table 44). This is again, of course, part of the environments item in the main analysis of variance. Finally to complete the analysis of this 3 X 3 table, we obtain a SS of 0.2674 for 2 X 2 = 4 df, giving a MS of 0.0669 for the interaction of the genetic differences among the three lines with the environmental differences among the three container types (I2′ in the lowest part of Table 44). The overall effect of container types (E2) gives a VR of = 0.1750/0.020 72 = 8.4, when compared with the error variance, showing significance at $P \triangleq 0.001$. The VR for the interaction of container types with lines (I2′) when compared with error is 3.2, which again is significant, P lying between 0.05 and 0.01. Evidently the three lines are not alike in their reactions to container type, although this interaction is smaller than the interaction with temperature (I1′), just as the overall effect of container (E2) is smaller than that of temperature (E1). We shall discuss the comparisons of temperature and container interactions further in the next Section.

As we have noted, this approach to genotype X environment interactions, through the analysis of variance, utilizes the formulation in $[d]$, e and g. The second approach to which we now turn, utilizes the alternative formulation which is represented in the simple use of S and W at the two temperatures by $[d]$, e_s and e_w. Now e_s and e_w measure the differences in S and W produced by the change in temperature. We could obviously introduce further parameters to represent the changes produced in S and W by the changes in culture containers. Altogether five orthogonal parameters would be required to specify the differences of chaeta number in S among the six environments and similarly five more for W. Now given that they are orthogonal to one another, as they must be if the specification is to be adequate, the five parameters for, say S, will make independent contributions to the SS, for of course 5 df, among the six chaeta numbers, one from each environment. Thus V_S, the variance of S over environments, will reflect the values of these five e_s parameters, and so the response of this genotype to the environmental changes. V_W, the variance of W over environments will similarly reflect the values of the five e_w parameters, and if the corresponding e_s and e_w parameters are not equal to one another, that is if there is genotype × environment interaction, V_S will not in general equal V_W. So we can detect the presence of genotype × environment interaction by comparing the variances of the different lines taken over environments.

This test is applied to the data of Table 40 including now the F_2 since the difference between its and the other error variance is of lesser importance in relation to this procedure. The results of the test are set out in Table 45, from which it is immediately apparent that the variances

TABLE 45

Variances over environments of S, W, F_1 and F_2 (Table 40)

	Over all environments df = 5	Over temperatures 1	Remainder 4
S	0.0508	0.0771	0.0443
W	0.6280	2.5220	0.1544
F_1	0.4723	2.0651	0.0742
F_2	0.3777	1.4211	0.1168

All entries are mean squares

are not alike. In particular $V_W = 0.628$ is much larger than $V_S = 0.0508$, giving with it a VR of $V_W/V_S = 12.4$ which for 5/5 df has a probability of $P < 0.02$, after doubling the P to allow for putting the larger variance V_W over the smaller V_S in the VR. There is no doubt about the significance of the interaction of these genotypes with the environments, as indeed we have already found using the analysis of variance. The comparison of V_W and V_S however, gives us further information not immediately available from the analysis of variance: since V_W is bigger than V_S, the W line must change more than S over environments - W is more sensitive to environmental change than S. Furthermore, V_{F1} and V_{F2} are significantly greater than V_S, although neither is significantly smaller than V_W. It would thus appear that both F_1 and F_2 are closer to their W parent than to S in their sensitivity to environmental change. We must however, still be a little cautious where the F_2 is concerned because of its markedly larger error variance (Table 40).

We can partition the changes over the six environments into that related to temperature for 1 df, and the rest involving types of culture container, for 4 df. In respect of W, the total SS over all six environments, is

$$SS_W = (19.63^2 + 19.34^2 \cdots + 17.61^2) - \tfrac{1}{6}(19.63 + 19.34 \cdots + 17.61)^2 = 3.1397$$

giving $V_W = 3.1397/5 = 0.6280$ as entered in Table 45.

The SS for the temperature difference is similarly

$$SS_{WT} = \tfrac{1}{3}[(19.63 + 19.34 + 19.34)^2 + (18.67 + 18.14 + 17.61)^2] - \tfrac{1}{6}(112.73)^2 = 2.5220$$

which corresponding as it does to 1 df, is also V_{WT}. The SS remaining, and corresponding to 4 df, is thus

$$SS_{WR} = 3.1397 - 2.5220 = 0.6177 \quad \text{giving} \quad V_{WR} = 0.1544.$$

These and similar results for S, F_1 and F_2 are included in Table 45.

In W; F_1 and F_2 the effect of temperature accounts for much of the major response to the environmental changes, the V_T significantly exceeding the V_R at the 0.01 level of probability in W and F_1, and exceeding at the 0.05 level in F_2. With S the effect of temperature is relatively much smaller, and although V_T is greater than V_R even in this case, it is not significantly so. The differences among the four lines for V_R are not significant, but again V_R is smallest for S, largest for W and intermediate for F_1 and F_2. There is thus at least a hint that the order of the four lines is basically the same for sensitivity to changes involving container type as for sensitivity to the temperature change. It should be observed, however, that although V_R is much smaller than V_T for all lines but S, it is on the other hand significantly greater than the relevant error variance (see Table 40) in W and F_1 and even in S it is approaching significance. Thus W and F_1, at least, do change with container, although to a smaller extent than with temperature, while S gives some appearance (even if it is not significant) of being less sensitive than the others to changes involving container type as well as to temperature. The high error variance of F_2 renders it relatively uniformative in the present connection; but even leaving it aside, the results from the analysis again pose the question of whether W and F_1 are just more sensitive to any environmental change than is S, or whether the differences in reaction to temperature and container changes, although both smaller in S than in the others, fail in fact to follow precisely the same relative patterns in all the lines. This is a question which we must now examine further.

27. The relation of *g* to *e*

So far our discussion of genotype × environment interaction has not required us to introduce measurements of the environment such as would allow the quantification of the environmental changes and hence the comparison, one with another, of different changes in their effects on the interaction. Such quantification of the changes is of course easy enough where temperature is altered; it can be measured in °C. The temperature change used in the experiment we have been discussing was 7° C, and if more than two temperatures had been used, the changes they represented could have been compared on this scale. The changes in container and the culture conditions which they imply, are however not so easily quantifiable: there is no obvious scale on which we can simultaneously represent the differences in size and shape of the containers themselves and the differences in food mass and supply of yeast. Furthermore, if we are to compare the differences in response of the two lines S and W to change of temperature with their differences in response to containers we need a single scale on which all the variations in environment can be represented. The only way in which we can achieve this is by seeking a biological measurement of the environment and its changes, that is by measuring the environment through its effects on the organisms themselves.

If for a moment we confine ourselves to the temperature difference and go back to the formulation in $[d]$, e and g as set out in Table 42, we see that the average of the S and W chaeta number at 18° C was $m + e$ and at 25° C was $m - e$. These averages are independent of both $[d]$ and g. They thus afford us a measure of the average or overall effect of the change in temperature. The bottom margin of Table 41 shows that the average at 18° C was $m + e = 39.89/2 = 19.945$, and at 25° C was $m - e = 19.410$, thus giving $e = \frac{1}{2}(19.945 - 19.410) = 0.2675$, as indeed we found earlier (Section 25).

Now if instead of taking the average, i.e. half the sum, of the S and W chaeta number at 18° C we had taken half their difference, we see from Table 42 that this will give us

$$\tfrac{1}{2}(\mathrm{S}-\mathrm{W}) = \tfrac{1}{2}\{(m+[d]+e+g)-(m-[d]+e-g)\} = [d]+g.$$

At 25° C we find similarly that $\frac{1}{2}(\mathrm{S}-\mathrm{W}) = [d]-g$. Then taking the data of Table 41, $[d]+g = \frac{1}{2}(20.45 - 19.44) = 0.505$ and $[d]-g = \frac{1}{2}(20.68 - 18.14) = 1.270$ giving $[d] = 0.8875$ and $g = -0.3825$ again as found earlier. Thus when the overall effect of the environment changes by $e = 0.2675$ the interaction changes by $g = -0.3825$. In other words the ratio of change in the interaction to that in the overall effect of the environment is $g/e = -0.3825/0.2675 = -1.4299$. So, given that there is a straight line relation between g and e (which with only two temperatures we cannot test and hence must be cautious in assuming) a change of temperature that produces an effect e in the average chaeta number given by these genotypes will then alter their difference by $-1.4299 \times 2e = 2.8598e$. It will do so by virtue of a change in S of $e+g = -0.4299e$ and a change in W of $e-g = 2.4299e$.

This treatment is readily extended to take all six environments into account. For convenience the environments have been numbered 1 to 6 where 1 is B at 18° C etc. as shown in Table 46. $\frac{1}{2}(\mathrm{S}+\mathrm{W})$ is then found

TABLE 46.

Relation of g to e in S and W

Environment	1 (18°C B)	2 (18°C Y)	3 (18°C U)	4 (25°C B)	5 (25°C Y)	6 (25°C U)	Mean
$\frac{1}{2}(\bar{S}+\bar{W}) = m+e$	20.105	19.925	19.800	19.555	19.535	19.135	19.6758 (= m)
e	0.4292	0.2492	0.1242	−0.1208	−0.1408	−0.5408	0
$\frac{1}{2}(\bar{S}-\bar{W}) = [d]+g$	0.475	0.585	0.460	0.885	1.395	1.525	0.8875 (= $[d]$)
g	−0.4125	−0.3025	−0.4275	−0.0025	0.5075	0.6375	0

SS(e) = 0.5886 SS(g) = 1.1084 SCP = −0.7214
$b = -1.2256$

Analysis of variance of g

Item	df	MS	VR	P
Regression	1	0.8842	85.3	v.s.
Remainder	4	0.0560	5.4	0.01-0.001
Error	48	0.0104		

VR for Regression/Remainder = 15.8 with P = 0.05-0.01

from the data of Table 40 as entered in Table 46. Each of these entries is $m + e$ where e_1 to e_6, from the six environments, sum to 0. m is the average of the six values and turns out to be 19.675 83, which on deducting from $\frac{1}{2}(S + W)$ gives the values of e_1 to e_6 as shown. (It should be noted that this value for m does not agree exactly with that found in Section 25, where only the temperatures were being considered, because the data of Table 41, although found from that of Table 40 were rounded off to the second decimal place for ease of calculation.) Next the six values of $\frac{1}{2}(S - W)$ are found. These are $[d] + g$, and since the six values of g must sum to 0, their average gives $[d] = 0.8875$, which on deducting from $\frac{1}{2}(S - W)$ leaves the six g's.

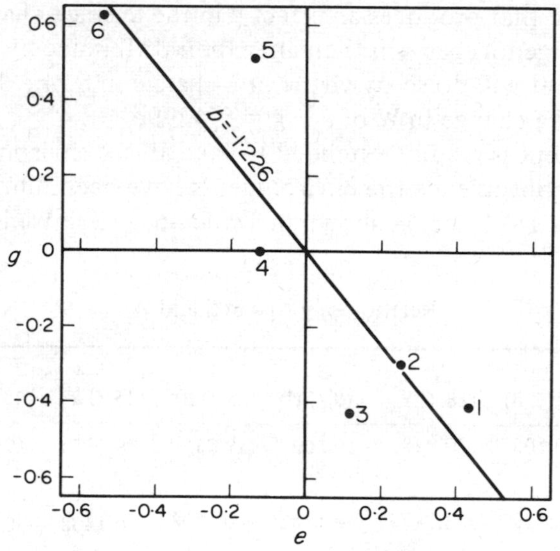

Fig. 16. The regression of g on e for sternopleural chaeta number in two lines of *Drosophila melanogaster* (S and W) raised in six environments (1 to 6). The regression line of g on e has a slope of −1.2256 which by being outside the range 1 to −1, shows that the two lines of flies respond in opposite directions to the relevant change in the environment (see also Fig. 17).

These six g's are plotted against their corresponding e's in Fig. 16 from which it is clear that there is a negative relation between g and e, g falling as e rises. We can test whether this relation is rectilinear by finding the regression of g on e. The calculation is shown at the foot of Table 46. The SS for e is found simply as $e_1^2 + e_2^2 \ldots e_6^2$, since the sum of the six e's must be 0. (It is nevertheless easier to find this SS as $(m + e_1)^2$

$+ (m + e_2)^2 \ldots + (m + e_6)^2 - \frac{1}{6}[(m + e_1) + (m + e_2) \cdots + (m + e_6)]^2$ as every $m + e$ is known exactly whereas all the e's involve recurring decimals.) Similarly $SS_{(g)} = g_1{}^2 + g_2{}^2 \cdots + g_6{}^2$ while the S.C.P. of g and e is $e_1 g_1 + e_2 g_2 \cdots e_6 g_6$. Then the linear regression coefficient of g on e is S.C.P./$SS_{(e)} = -0.7214/0.5886 = -1.2256$. The analysis of variance of g is carried out in the customary way, the SS for regression being $(-0.7214)^2/0.5886 = 0.8842$ which on subtracting from $SS_{(g)}$ leaves $1.1084 - 0.8842 = 0.2242$ as the SS remaining. Since there are six environments each yielding an observation, there will be 5 df of which 1 is taken up by the regression itself leaving 4 df for variation of the points round the regression line, so giving as the remainder MS $0.2242/4 = 0.0560$.

Each g value is derived from that of the difference between an S observation and a W observation. Each observation is subject to an error variance of 0.020 72 and the difference between two of them will have an error variance twice this value. Half the difference will have an error variance of one-quarter that of the difference itself. Thus g will be subject to a variance of $\frac{1}{4} \times 2 \times 0.020\,72 = 0.010\,36$. When tested against this estimate of error the remainder MS gives a VR of 5.41 for 4 and 48 df and this has a P between 0.01 and 0.001. The departures from the linear regression are thus significant. At the same time the regression MS tested against the remainder MS yields a VR of 15.78 which for 1 and 4 df has $P = 0.02 - 0.01$. Thus, despite the variation round the line, there can be no doubt of the linear component in the regression of g on e.

This linear component must reflect the relation between g and e which we have already seen to be produced by the temperature changes. This relation plays a dominant part in producing the regression line because the effect of temperature in changing e is greater than the effects of the changes in culture container as a glance at Fig. 16 will confirm. The significant variation about the regression line reflects the consequences of the changes in container, which must thus produce interactions, g, not related in the same way to the overall effects, e, as those brought about by the alteration of temperature. Thus the relative responses of the two genotypes to change in culture container cannot be following the same pattern as their relative responses to change in temperature. It is therefore necessary to specify the type of environmental change before we can discuss the relative sensitivities of the two genotypes to it.

The plot of g against e in Fig. 16 brings out in a clear and simple way the relation between these two quantities. It shows us, however, nothing of the sensitivities to environmental change of the individual genotypes

S and W. A more informative, albeit somewhat more complex, picture can be obtained in a slightly different way. If we deduct m from the values given by S in the six environments we are left with $[d] + e_1 + g_1$, $[d] + e_2 + g_2$, etc., which may of course be written in the alternative formulation as $[d] + e_{s1}$ $[d] + e_{s2}$, etc. Similarly deducting m from the values given by W leaves $-[d] + e_1 - g_1$, $-[d] + e_2 - g_2$, etc. which may also be rewritten as $-[d] + e_{w1}$, $-[d] + e_{w2}$, etc. The values so obtained for the two genotypes are set out in Table 47 and are plotted against e in Fig. 17. The table also gives the linear regression coefficients, b, of $S - m$ and $W - m$ on e, and the regression lines themselves are shown on the figure.

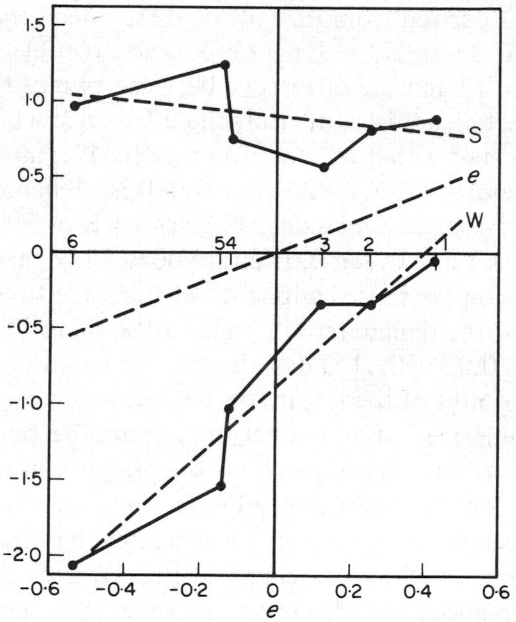

Fig. 17. The sensitivity diagram for sternopleural chaeta number in S and W. The deviations of S and W from the mid-parent, m, for each environment are plotted (ordinate) against e (abscissa). The six environments are denoted by the numbers 1 to 6. The outer broken lines are the best fitting regression lines of $S - m$ and $W - m$ on e. The mean of $S - m$ and $W - m$ is e for each environment, and the central broken line derived from these means is thus the regression of e on e and must have a slope of 1. The diagram makes clear that W is more sensitive than S to change in the environment and that the two change in opposite directions.

TABLE 47.

Sensitivity to environmental change in S, W, their F_1 and F_2

	Environment						Mean	b
	1	2	3	4	5	6		
$\overline{S}-m$ $(=[d]+e_s)$	0.9042	0.8342	0.5842	0.7642	1.2542	0.9842	0.8875	−0.226
$\overline{W}-m$ $(=[-d]+e_w)$	−0.0458	−0.3358	−0.3358	−1.0058	−1.5358	−2.0658	−0.8875	2.226
e $(=\frac{1}{2}[e_s+e_w])$	0.4292	0.2492	0.1242	−0.1208	−0.1408	−0.5408	0	1.000
$\overline{F}_1-m$ $(=[h]+e_h)$	0.3042	0.3342	0.4842	−0.4558	−0.7458	−1.1958	−0.2125	1.836
$\overline{F}_2-m$ $(=\frac{1}{2}[h]+\frac{1}{2}e_h)$	0.5142	0:1842	0.0742	−0.2258	−0.9958	−0.9258	−0.2292	1.603

The first point to note is that the means of $\bar{S} - m$ and $\bar{W} - m$ are $[d]$ and $-[d]$ respectively. The regression line for $\bar{S} - m$ must thus cut the ordinate of the graph at $[d]$, and the regression line for $\bar{W} - m$ cuts it at $-[d]$. These two points must be equally spaced above and below the origin, as will be seen from the figure. Next, the slope of the $\bar{S} - m$ regression line measures the rate of change of $e_s = e + g$ on e: in other words it measures the sensitivity of S to change in the environment. Equally the slope of the $W - m$ line measures the sensitivity of W to change in the environment, and clearly this is much greater than the sensitivity of S, which in so far as it changes at all does so in the opposite direction. Now the slope of the S line depends on the change in $e_s = e + g$ on e, while that of the W line depends on $e_w = e - g$. The slope observed for the S line is -0.226 while that of the W line is 2.226. The regression of e on e, which is also shown in the figure, will obviously have a slope of 1. Thus the slope of the S line departs from that of e by $-0.226 - 1.000 = -1.226$, while that of the W line departs by $2.226 - 1.000 = 1.226$. So the interaction of genotype with environment is responsible for a slope of -1.226 in S and 1.226 in W - values which are equal in magnitude but opposite in sign as indeed they must be since e is found from the mean of S and W in each environment, with the S line reflecting the change of g and the W line that of $-g$. We may also note that -1.226 measuring the contribution of g to the slope of the S line equals the slope we have already found in a different way for the regression of g on e (Table 46 and Fig. 16) as indeed it must. Table 47 and Fig. 17 thus give us all the information that we were able to obtain from Table 46 and Fig. 16 and more besides.

This analysis of the genotype × environment interaction is made possible only by using the chaeta numbers displayed by S and W in the different environments to provide their own biological measurement of the environments and so to quantify the overall effects of various changes of environments. The biological measure, e, has allowed us to quantify the consequences of the changes in culture condition as well as those of change in temperature and show both on the same scale. In doing so it has enabled us to compare the patterns of response to temperature and culture condition and show that they are not the same. A further advantage, although not one that is brought out by our present data, is that g may display a rectilinear relation to environmental change measured by e, even where it fails to do so when the environment is measured in other and perhaps more obvious ways. As an example of this, two strains of the fungus *Schizophyllum commune* (Jinks and Connolly, 1973) when grown in a series of nine environments differing by temperature, display interactions which when quantified by g are related in a curvilinear manner to temperature itself. But when the temperature is replaced by the biological measure e the relation of g to the environmental change becomes rectilinear, as is shown in Fig. 18.

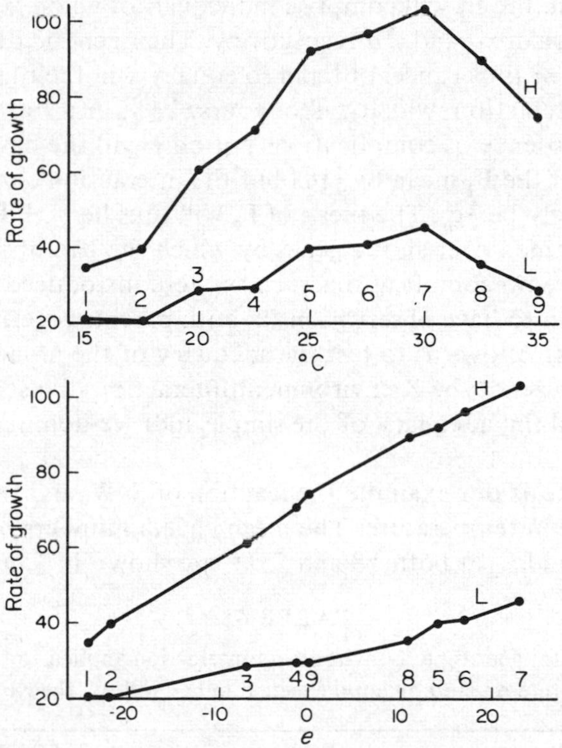

Fig. 18. The effect of temperature on growth rate (in mm per nine days) of a slow (L) and a fast (H) growing strain of *Schizophyllum commune.* The upper graph shows growth rates plotted against temperature, and the lower graph shows it plotted against e, the biological measure of the nine environments. The nine temperatures are denoted by the numbers 1 to 9, which thus relate corresponding points on the two graphs.

28. Crosses between inbred lines

As we have seen, the F_1 and F_2 generations were raised in the six environments in addition to the S and W parental lines. Now the departure of the mean chaeta number of F_1 from m, the mid-parent value, is $[h]$ and that of the F_2 mean is $\frac{1}{2}[h]$. So whereas the interaction of the parental genetic difference is the interaction of $[d]$ with e, the interaction of the F_1 genotype, and with it any interaction of the F_2 mean, will depend on the interaction of $[h]$ with e. We must therefore, distinguish between two g parameters, g_d measuring the interaction of $[d]$ and e, and g_h measuring the interaction of $[h]$ with e. The mean phenotype of the F_1 thus becomes $m + [h] + e + g_h$. In respect of any single gene difference A-a, the F_2 will comprise individuals of whom $\frac{1}{4}$ will be AA, and $\frac{1}{4}$ aa, which are d and $-d$ respectively. Their genetic deviation of d and $-d$ from m thus cancel out and so equally will their genotype X environment interaction, which will of course be g_d and $-g_d$ respectively. Thus in the absence of complications not only will the basic genetic component of the F_2 mean be $\frac{1}{2}[h]$ but the interaction component will correspondingly be $\frac{1}{2}g_h$. The mean of F_2 will thus be $m + \frac{1}{2}[h] + e + \frac{1}{2}g_h$ when summed over all the genes by which the parents differ, provided there are no complications such as those introduced by non-allelic interaction. So to have observations in both parents together with their F_1 and F_2 mean allows us to test the adequacy of the model we have developed for genotype X environment interaction in just the same way that we tested the adequacy of the simple additive-dominance model in Section 9.

We will take as our example the reaction of S, W, and their derivatives to the change in temperature. The mean chaeta numbers of the parents and their F_1 and F_2 at both 18 and 25° C are shown in Table 48, together

TABLE 48.

The model for genotype X environment interaction applied to the effect of temperature on sternopleural chaeta number in S, W, their F_1 and F_2

	Temp.	Weight	m	$[d]$	$[h]$	e	g_d	g_h	Observed	Expected
S	18	144.783	1	1	0	1	1	0	20.450	20.447
	25	144.783	1	1	0	−1	−1	0	20.667	20.668
W	18	144.783	1	−1	0	1	−1	0	19.437	19.434
	25	144.783	1	−1	0	−1	1	0	18.140	18.131
F_1	18	290.360	1	0	1	1	0	1	20.050	20.047
	25	290.360	1	0	1	−1	0	−1	18.877	18.868
F_2	18	29.463	1	0	$\frac{1}{2}$	1	0	$\frac{1}{2}$	19.933	19.994
	25	29.463	1	0	$\frac{1}{2}$	−1	0	$-\frac{1}{2}$	18.960	19.134

$\chi^2_{[2]} = 1.046 \quad P = 0.7 - 0.5$

with their structures in terms of the six parameters, m, $[d]$, $[h]$, e, g_d, g_h and also the weights attached to each observed chaeta number in the analysis. The weights come of course from the variances given in Table 40. Since each temperature mean is found by averaging three observations, B, Y and U, at that temperature, its variance in for example the parent lines will be $0.020\,72 \div 3 = 0.006\,907$ and the weight is $1/0.006\,907 = 144.78$. The six weighted least squares equations of estimation for the parameters are then obtained in a manner exactly analogous to that used in Section 9 and turns out to be, in matrix form,

$$\underset{\mathbf{J}}{\begin{bmatrix} 1218.777 & 0 & 610.183 & 0 & 0 & 0 \\ 0 & 579.131 & 0 & 0 & 0 & 0 \\ 610.183 & 0 & 595.451 & 0 & 0 & 0 \\ 0 & 0 & 0 & 1218.777 & 0 & 610.183 \\ 0 & 0 & 0 & 0 & 579.131 & 0 \\ 0 & 0 & 0 & 610.183 & 0 & 595.451 \end{bmatrix}} \underset{\hat{\mathbf{M}}}{\begin{bmatrix} \hat{m} \\ [\hat{d}] \\ [\hat{h}] \\ \hat{e} \\ \hat{g}_d \\ \hat{g}_h \end{bmatrix}} = \underset{\mathbf{S}.}{\begin{bmatrix} 23\,843.546 \\ 513.979 \\ 11\,875.702 \\ 524.284 \\ -220.553 \\ 355.028 \end{bmatrix}}$$

Inversion of the **J** matrix thus enables us to write

$$\underset{\hat{\mathbf{M}}}{\begin{bmatrix} \hat{m} \\ [\hat{d}] \\ [\hat{h}] \\ \hat{e} \\ \hat{g}_d \\ \hat{g}_h \end{bmatrix}} = \underset{\mathbf{J}^{-1}}{\begin{bmatrix} 0.001\,684\,9 & 0 & -0.001\,726\,6 & 0 & 0 & 0 \\ 0 & 0.001\,726\,7 & 0 & 0 & 0 & 0 \\ -0.001\,726\,6 & 0 & 0.003\,448\,7 & 0 & 0 & 0 \\ 0 & 0 & 0 & 0.001\,684\,9 & 0 & -0.001\,726\,6 \\ 0 & 0 & 0 & 0 & 0.001\,726\,7 & 0 \\ 0 & 0 & 0 & -0.001\,726\,6 & 0 & 0.003\,448\,7 \end{bmatrix}} \underset{\mathbf{S}}{\begin{bmatrix} 23\,843.546 \\ 513.979 \\ 11\,875.702 \\ 524.284 \\ -220.553 \\ 355.028 \end{bmatrix}}$$

from which we find

$$\begin{aligned} \hat{m} &= 19.6799 \pm 0.0411 \\ [\hat{d}] &= 0.8875 \pm 0.0416 \\ [\hat{h}] &= -0.2125 \pm 0.0587 \\ \hat{e} &= 0.2704 \pm 0.0411 \\ \hat{g}_d &= -0.3808 \pm 0.0416 \\ \hat{g}_h &= 0.3192 \pm 0.0587. \end{aligned}$$

the standard errors being obtained as the square roots of the entries along the leading diagonal of the variance-covariance matrix $\mathbf{J}^{-1}$. All the estimates are significant and no parameter is redundant therefore.

The estimates allow us to calculate expectations for the mean chaeta number of the S, W, F_1 and F_2 at each temperature as shown in the last column of Table 48. Then squaring the differences between observed and expected means, multiplying each squared difference by the corresponding weight and summing over all eight observations gives $\chi^2_{[2]} = 1.046$, there being 2 df because six parameters have been estimated from the eight observations. This $\chi^2_{[2]}$ has a probability lying between 0.7 and 0.5, indicating that so far as these data go the model is fully adequate to account for the observations: there are no grounds for suspecting complications such as non-allelic interaction.

We should note, however, that a more sensitive test would have been possible if more generations, notably the two back-crosses, had been included in the experiment and if more replicates had been raised of the F_2 to reduce the variance of its mean.

Comparing the estimates of the parameters with their contributions to the eight observations shows that:

(i) $[\hat{d}]$ is positive because S has a larger mean number of chaetae than W.

(ii) $[\hat{h}]$ is negative because the F_1 and F_2 are nearer to W, the $-[d]$ parent, than to S which has $[d]$.

(iii) $\hat{e}$ is positive because the average chaeta number is higher at 18 than at 25° C.

(iv) $\hat{g}_d$ is negative because the difference between S and W decreases as the overall chaeta number rises from 25 to 18° C.

(v) $\hat{g}_h$ has the opposite sign to $[\hat{h}]$ because dominance decreases as the chaeta number rises from 25 to 18° C.

These points become clearer, if, having satisfied ourselves that on the one hand the model is adequate while on the other it contains no redundant parameters, we set out the analysis and its results in a different way. If we concentrate on the data from a single environment we have no information about the effects of environmental change. The four observations from one environment can therefore be accounted for by estimating only, m, $[d]$ and $[h]$, the estimates so obtained being of course applicable only to that environment. Proceeding in this way, one environment at a time, we obtain two estimates each of m, $[d]$ and $[h]$ thus:

	25° C	18° C	s.d.
m	19.3995	19.9403	± 0.0581
$[d]$	1.2683	0.5067	± 0.0588
$[h]$	−0.5316	0.1067	± 0.0831
$\chi^2_{[1]}$	0.934	0.112	

There are two χ^2's one from each environment and each having $4 - 3 = 1$ df. Neither is significant and the model is thus adequate at both environments.

Now $\hat{m}$ in the combined analysis is a combination of the two m's from the separate environments while $\hat{e}$ is a measure of the difference between the two separate m's. Similarly the combined $[\hat{d}]$ is a compound of the two separate $[d]$'s, while $\hat{g}_d$ depends on their difference; and the combined $[\hat{h}]$ is a compound of the two separate $[h]$'s while $\hat{g}_h$ depends on their difference. In the present case, where the variances of corresponding observations are equal in the two environments, $\hat{m}$ is in fact the simple average of m_{18} and m_{25}, while $\hat{e}$ is half their difference, i.e. is $\frac{1}{2}(m_{18} - m_{25})$. Similarly $[\hat{d}] = \frac{1}{2}([d]_{18} + [d]_{25})$ and $\hat{g}_d = \frac{1}{2}([d]_{18} - [d]_{25})$ while $[\hat{h}] = \frac{1}{2}([h]_{18} + [h]_{25})$ with $\hat{g}_h = \frac{1}{2}([h]_{18} - [h]_{25})$. The interpretation and implication of the estimates of the six parameters from the combined analysis of the results from the two environments are now clear. $[d]$ falls as m rises. Hence $[d]$ and e are moving in opposite directions, and g_d is thus negative. Similarly, while $[h]$ is preponderantly negative, it is rising as e rises and g_h is thus positive. A further point is brought out well by the present estimates. At 25° C $[h]$ is significantly negative, giving a ratio $[h]/[d] = -0.43$. At 18° C $[h]$ is positive but it does not differ significantly from 0, although it obviously does differ significantly from $[h]_{25}$. The ratio $[h]/[d] = 0.19$. Thus the dominance, or to be more precise the potence of the W genotype over the S changes markedly with the environment: the value of $[h]$ as indeed that of $[d]$ also, is not unconditional. This is of course another way of saying that the interaction between genotypes and environments affects dominance as well as additive variation.

One last point remains to be made about these results. The rate of change of g_d on e is $-0.3808/0.2704 = -1.4085$ which agrees with our estimate of -1.4299 obtained in the previous section from consideration of S and W alone. Since S departs from m by $[d]$ its interaction with the temperature change will thus be $-1.4085e$ but with W the deviation from m is $-[d]$ and the interaction is thus $-g_d = 1.4085e$. The rate of change of $\hat{g}_h$ on $\hat{e}$ is $0.3192/0.2704 = 1.1804$. Thus the reaction to temperature of the heterozygote is not only much nearer to that of W than it is to that of S - it is in fact approaching quite closely in value to that of W. Clearly W is dominant to S in its genotype × environment interaction: indeed its dominance in respect of the interaction is even greater than in respect of overall chaeta number.

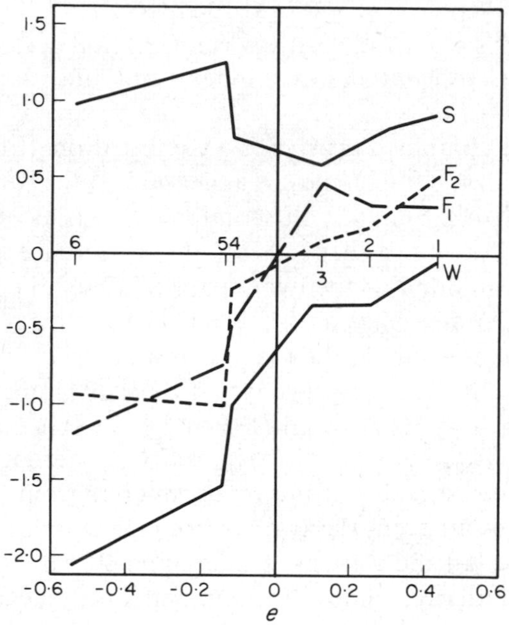

Fig. 19. The sensitivity diagram for sternopleural chaetas in the S and W lines of *Drosophila*, together with their F_1 and F_2. F_1 and F_2 follow the response pattern of W more than that of S, thus indicating the dominance of the relevant genes in W.

This is well seen from Fig. 19 which is a sensitivity diagram similar to that already presented for the parent lines in Fig. 17 but which now includes F_1 and F_2 as well. Although only the reaction to the temperature change was taken into account in the foregoing analysis, the figure shows the behaviour in all six environments. $\bar{F}_1 - m$ and $\bar{F}_2 - m$ are set out for all the environments in the lower part of Table 47, from which Fig. 19 has been drawn.

The regression lines for the two parents, F_1 and F_2 have been omitted from the figure, in the interests of clarity; but the regression coefficients (b) are shown in the right-hand column of Table 47. The expectations of $\bar{S} - m$, $\bar{W} - m$, $\bar{F}_1 - m$ and $\bar{F}_2 - m$ are shown on the left of the table from which it will be seen that just as b for S and W are the rates of change on e of $e + g_d$ and $e - g_d$ respectively, that of F_1 and F_2 are the rates of change of $e + g_h$ and $e + \frac{1}{2}g_h$. Since the rate of change of e on e is obviously 1, the rate of change (i.e. the regression) of g_h on e is $b_{F1} - 1 = 1.836 - 1 = 0.836$. We can then predict that the regression of $\frac{1}{2}g_h$ on e will be $0.836 \div 2 = 0.418$, which will give $0.418 + 1 = 1.418$ as the expected regression for the F_2. The observed regression is 1.603 which as expected is lower than that for the F_1. It is higher than the expectation but not significantly so.

The figure brings out very clearly, as do also the regression coefficients, the close similarity of F_1 and F_2 to W in their patterns of sensitivity to environmental change: in fact, the W pattern shows a high degree of dominance over that of S. In average chaetae number, on the other hand, although W is again dominant over S, the dominance is less, with the F_1 and F_2 means departing more in the direction of S. The results of this difference in dominance in respect of sensitivity and mean chaeta number is that in environments 1, 2 and 3 (the three at 18° C) the F_1 and F_2 are very close to half-way between W and S in chaeta number, i.e. show little or no dominance, while at environments 4, 5 and 6 (at 25° C) they are much closer to W than S. The dominance relations of W and S will thus depend on the environment in which they are measured. The diagram thus shows us both how and why the estimate of dominance can change, and change drastically, with alteration of the environment. It has also shown us the value of investigating sensitivity to environmental change as a character in its own right.

29. Variance of F_2

So far we have been considering the situation where the environments are defined and hence distinguishable from one another. The expression of the different genotypes can then be observed in each environment and the changes of expression related directly to change from one environment to another. The analysis is thus essentially one of components of means. Frequently, however, the environments are not so definable and unambiguously distinguishable. Thus, for example, the results from plants grown on distinct blocks in an experimental field can be handled by the methods we have been discussing because although we cannot specify the chemical or physical differences between the environments associated with the different blocks we can at least distinguish unambiguously the plants grown in the environment of block 1 from those grown in the environment of block 2 and so on. At the same time we would expect there to be similar, if smaller, differences between the environments in different parts of a single block, with the consequence that genotype × environment interaction must be affecting the results from plants from the same block, although these will not be assignable to environments identifiable as contrasting in the way possible where the comparison is between blocks. Where this is the case we must proceed in a different way, relying on variances rather than means for recognizing and analysing the interaction.

Let us consider a single gene difference on the one hand and a single environmental difference on the other. We assume that no matter what its genotype each individual has an equal chance of occurring in each of the two types of environment. Since the environments are not unambiguously distinguishable from one another, this condition must in practice generally require that the individuals irrespective of genotype are distributed at random over the range of environments present. Table 49 sets out the phenotypes expected from the three genotypes in the two

TABLE 49.

Contribution of $g \times e$ interaction to variances of parents, F_1 and F_2 over two environments

		Environment		Overall	
		1	2	Mean	Variance
Parent 1	AA	$d+e+g_d$	$d-e-g_d$	d	$(e+g_d)^2$
F_1	Aa	$h+e+g_h$	$h-e-g_h$	h	$(e+g_h)^2$
Parent 2	aa	$-d+e-g_d$	$-d-e+g_d$	$-d$	$(e-g_d)^2$
F_2 mean		$\frac{1}{2}h+e+\frac{1}{2}g_h$	$\frac{1}{2}h-e-\frac{1}{2}g_h$	$\frac{1}{2}h$	$\frac{1}{2}d^2+\frac{1}{2}g_d^2+\frac{1}{4}h^2$ $+\frac{1}{4}g_h^2+(e+\frac{1}{2}g_h)^2$

classes of environment, expressed in terms of their deviation from the mid-parent, m. Taking the three genotypes individually we see that their expressions, averaged over the two environments, are, d, h and $-d$ respectively. In other words the only means available tell us nothing about the interaction. Their variances however, do so. In the absence of interaction they will all be equal but with interaction present they will no longer be equal. They become $(e+g_d)^2$, $(e+g_h)^2$ and $(e-g_d)^2$ for AA,

Aa and aa respectively. Extending this to any number of environmental differences in respect of which the genotypes are distributed at random, we can see that even where the different environments are not directly distinguishable, genotype X environment interaction can be detected by differences in the variances of the phenotypes produced by the different genotypes. This is effectively the same test that we were using in Table 45, although in that case the environments were distinguishable from one another.

Turning to the F_2, where the environments are distinguishable and each individual hence assignable to its environment, the means of the F_2 in the two environments differ by $2(e + \frac{1}{2}g_h)$, one of them being $\frac{1}{2}h + e + \frac{1}{2}g_h$ and the other $\frac{1}{2}h - e - \frac{1}{2}g_h$. V_{1F2} also differs in the two environments. It can still be represented in the form $\frac{1}{2}D + \frac{1}{4}H$ but the definition of D and H change with the environment, the gene contributing $(d + g_d)^2$ and $(h + g_h)^2$ to D and H respectively in one environment and $(d - g_d)^2$ and $(h - g_h)^2$ in the other. We are in fact elaborating the simple definition of d and h that we use in a single environment to take into account the interaction of the gene with its environment.

Where the environments are not distinguishable and we must therefore take the mean and variance of the F_2 as a whole, the contribution of the gene-pair to the mean becomes $\frac{1}{2}h$, which is of course the average of the means in the two separate environments, $\frac{1}{2}h + e + \frac{1}{2}g_h$ and $\frac{1}{2}h - e - \frac{1}{2}g_h$. As with parents and F_1, the overall mean gives no information about the interaction. But again as in the earlier case the variance does reflect the interaction, being $V_{1F2} = \frac{1}{2}d^2 + \frac{1}{2}g_d^2 + \frac{1}{4}h^2 + \frac{1}{4}g_h^2 + (e + \frac{1}{2}g_h)^2$. Now if we use the variances of the two parental homozygotes and their F_1 to provide an estimate of non-heritable variation as we have done in earlier chapters and combine them in the F_2 ratio itself, i.e. by finding $\frac{1}{4}V_{AA} + \frac{1}{2}V_{Aa} + \frac{1}{4}V_{aa}$, our estimate becomes

$$\tfrac{1}{4}(e + g_d)^2 + \tfrac{1}{2}(e + g_h)^2 + \tfrac{1}{4}(e - g_d)^2 = \tfrac{1}{2}g_d^2 + \tfrac{1}{4}g_h^2 + (e + \tfrac{1}{2}g_h)^2.$$

Deducting this from V_{1F2} to estimate the gene's contribution to the heritable component of the F_2 variance leaves us with

$$\begin{aligned} {}_HV_{1F2} &= \tfrac{1}{2}d^2 + \tfrac{1}{2}g_d^2 + \tfrac{1}{4}h^2 + \tfrac{1}{4}g_h^2 + (e + \tfrac{1}{2}g_h)^2 \\ &\quad - \tfrac{1}{2}g_d^2 - \tfrac{1}{4}g_h^2 - (e + \tfrac{1}{2}g_h)^2 = \tfrac{1}{2}d^2 + \tfrac{1}{4}h^2 \end{aligned}$$

which is the same as is found in the absence of genotype X environment interaction.

The result is not difficult to generalize for more than two environments. Consider t environments 1 to t, among which parents, F_1 and F_2 are distributed at random, the probability of any individual falling into a given environment being $1/t$ i.e. equal for all environments. (Note that if one type of environment is more common than another, it can be accommodated in the formulation by letting an appropriate number of the t environments all be of this kind.) Each environment has its own e, g_d and g_h, those in environment 1 being e_1, g_{d1} and g_{h1} etc., where $S(e) = 0$, $S(g_d) = 0$ and $S(g_h) = 0$. The phenotypes of parents and F_1, and the mean phenotype of F_2 in each environment will be as set out in Table 50.

TABLE 50.

Variances of parents, F_1 and F_2 over t environments

		Environment	Overall	
		$1 \cdots\cdots\cdots t$	Mean	Variance
Parent 1	AA	$d+e_1+g_{d1}\cdots d+e_t+g_{dt}$	d	$S(e+g_d)^2$
F_1	Aa	$h+e_1+g_{h1}\cdots h+e_t+g_{ht}$	h	$S(e+g_h)^2$
Parent 2	aa	$-d+e_1-g_{d1}\cdots -d+e_t-g_{dt}$	$-d$	$S(e-g_d)^2$
F_2 mean		$\frac{1}{2}h+e_1+\frac{1}{2}g_{h1}\cdots\frac{1}{2}h+e_t+\frac{1}{2}g_{ht}$	$\frac{1}{2}h$	$\frac{1}{2}d^2+\frac{1}{2}Sg_d^2+\frac{1}{4}h^2$ $+\frac{1}{4}Sg_h^2+S(e+\frac{1}{2}g_h)^2$

Then taken over all environments, the means of the parents are d and $-d$ respectively, that of F_1 is h and that of F_2 is $\frac{1}{2}h$. The variance of the AA parent will be $S(e+g_d)^2$ which will also be $V(e+g_d)$ since with each environment carrying $1/t$ of the individuals the SS will also be the MS. The variances of the other parent, F_1 and F_2 are similarly shown on the right-hand column of the table. Now when the parental and F_1 variances are combined in the F_2 proportions they give $\frac{1}{2}S(g_d)^2+\frac{1}{4}S(g_h)^2+S(e+\frac{1}{2}g_h)^2$ and subtracting this from V_{1F2} gives the heritable component due to the gene A-a as

$$
{}_HV_{1F2} = \tfrac{1}{2}d^2+\tfrac{1}{2}S(g_d)^2+\tfrac{1}{4}h^2+\tfrac{1}{4}S(g_h)^2+S(e+\tfrac{1}{2}g_h)^2 \\ -\tfrac{1}{2}S(g_d)^2-\tfrac{1}{4}S(g_h)^2-S(e+\tfrac{1}{2}g_h)^2
$$

just as we found earlier in the simpler case of two environments.

The extension to more than one gene difference however, brings in a new problem. This is simply illustrated by the case of two gene differences, A-a and B-b, in two environments. It is easy to show that the variances of the four possible homozygotes will be

$$V_{AABB} = (e + g_{da} + g_{db})^2;\ V_{AAbb} = (e + g_{da} - g_{db})^2;$$
$$V_{aaBB} = (e - g_{da} + g_{db})^2;\ V_{aabb} = (e - g_{da} - g_{db})^2.$$

Thus if we use AABB and aabb as the parents from whose cross the F_2 is raised, the average of their variances will clearly be $(g_{da} + g_{db})^2 + e^2$ while with the alternative pair of parents, AAbb and aaBB, it will be $(g_{da} - g_{db})^2 + e^2$. The variance of F_1 will be $(e + g_{ha} + g_{hb})^2$ in both cases, so combining parents and F_1 variances in the F_2 proportions will give

$$\tfrac{1}{2}(g_{da} + g_{db})^2 + \tfrac{1}{4}(g_{ha} + g_{hb})^2 + (e + \tfrac{1}{2}g_{ha} + \tfrac{1}{2}g_{hb})^2$$

with the cross AABB × aabb

and

$$\tfrac{1}{2}(g_{da} - g_{db})^2 + \tfrac{1}{4}(g_{ha} + g_{hb})^2 + (e + \tfrac{1}{2}g_{ha} + \tfrac{1}{2}g_{hb})^2$$

with the cross AAbb × aaBB.

Now in the absence of linkage the composition of the F_2 will be the same from both crosses and its variance in respect of these two gene pairs will be

$$V_{1F2} = \tfrac{1}{2}d_a^2 + \tfrac{1}{2}d_b^2 + \tfrac{1}{2}g_{da}^2 + \tfrac{1}{2}g_{db}^2 + \tfrac{1}{4}h_a^2 + \tfrac{1}{4}h_b^2 + \tfrac{1}{4}g_{ha}^2 + \tfrac{1}{4}g_{hb}^2 + (e + \tfrac{1}{2}g_{ha} + \tfrac{1}{2}g_{hb})^2$$

which, after deducting the variances of parents and F_1 combined in F_2 proportions leaves

$${}_HV_{1F2} = \tfrac{1}{2}d_a^2 + \tfrac{1}{2}d_b^2 \mp g_{da}g_{db} + \tfrac{1}{4}h_a^2 + \tfrac{1}{4}h_b^2 - \tfrac{1}{2}g_{ha}g_{hb}$$

the term in $g_{da}g_{db}$ being negative where the cross was AABB × aabb and positive where it was AAbb × aaBB. The estimate of the basic genetical component of the variation is thus not free from the effects of the environmental interaction where two or more genes are involved. These residual effects depend on cross-product terms of the kinds $g_{da}g_{db}$ and $g_{ha}g_{hb}$ and as the number of genes rises the number of such terms rises relative to the number of squared terms of the kinds $g_{da}^2, g_{db}^2, g_{ha}^2, g_{hb}^2$ which are eliminated. The residual effects are therefore prospectively the more troublesome as the number of genes in the system increases.

In the case of the g_d terms the residual effects could be eliminated if all the homozygotes (four with two gene pairs) were available for their variances to be compounded in finding the correction to be deducted from V_{1F2}, but this will seldom be possible. The signs of the terms in $g_{d.}\, g_{d.}$ will however, depend not only on the intrinsic signs of the individual g_d's but also on whether the relevant genes are associated or disper in the parental homozygotes. If the genes are suitably dispersed between the parents the net result could be that on summing over all pairs of gene differences the aggregate $S(g_{da}\, g_{db})$ was negligible. The estimate of $D = S(d^2)$ would then not be greatly affected by the covariance of the interactions. The sign of the terms in $g_{h.}\, g_{h.}$ on the other hand, depends only on the intrinsic signs of the individual g_h's. Unless therefore there is an approach to equality in the number of positive and negative g's, the aggregate $S(g_{ha}\, g_{hb})$ cannot be expected to become negligible.

Similar terms in $S(g_{d.}\, g_d)$ and $S(g_{h.}\, g_h)$ will be associated with the contributions made by $D = S(d^2)$ and $H = S(h^2)$ respectively in the variance derived from later generations such as F_3. The relative size of the contributions made by these terms will depend not only on the variance in question, whether for example it is V_{1F3} or V_{2F3}, but also on the detailed design of the experiment from which the variances are estimated. The presence of genotype × environment interaction is, however, always liable to introduce bias into the estimates of D and H, the amount of bias depending on the extent to which the different $g_d\, g_d$ and $g_h\, g_h$ items balance out in $S(g_d \cdot g_d)$ and $S(g_h \cdot g_h)$ respectively. Thus, wherever differences in the variances of the two parental lines and the F_1 suggest sizeable interaction components of variation, we must treat the estimates of D and H with corresponding caution.

7

Randomly breeding populations

30. The components of variation

So far we have been concerned with the analysis of data obtained from true-breeding lines and the descendants of crosses made between them. Following such a cross, a multiplicity of generations and types of family can be raised experimentally - a multiplicity limited only by the biological properties of the material (whether, for example, it can be selfed as well as crossed, whether individuals can be kept alive for crossing to their own offspring and so on) and by the time and facilities available for the experimental programme. Each generation and type of family will have its own mean and variance, and its own covariances with other related families. Thus a large number of statistics can be obtained from which we can estimate the genetical and environmental components of both means and variances. The specification of these components of variation is simpler because by starting with true-breeding lines we can, in the absence of selective elimination, specify the relative frequencies of the types of zygotes and gametes that we expect in and from any given type of family.

When however we turn from the descendants of crosses among true-breeding lines to consider genetically heterogeneous populations of unspecified constitution, not only is the situation more complex, but the range of statistics available from the populations is commonly more limited. We can of course ascertain the mean and variance of the population itself; but given that it is in equilibrium and that non-heritable effects are not changing, these will be the same within sampling variation from one generation to the next. We can also compare the variation within families raised from pairs of parents with the variation between them, and we can look at the covariation between individuals of different genetical relationships, such as parents and offspring, full-sibs, half-sibs, first cousins and so on, provided we can recognize individuals with these relationships. Our analysis will thus depend on differences in second degree statistics rather than means and we shall not in general have the direct estimates of non-heritable variation that are provided by the variation of homozygous lines and their F_1s in the experiments we have discussed in earlier chapters.

Let us consider the gene pair A-a in a population in which mating is at random, the frequency of allele A being u_a and that of allele a being $v_a = 1 - u_a$. The incidence of the three genotypes in respect of this gene pair will then be AA u_a^2; Aa $2u_a v_a$; aa v_a^2. AA and aa deviate by d_a and $-d_a$ respectively from the mid-parent and Aa by h_a. Then in respect of this gene pair, the population mean will be $u_a^2 d_a + 2u_a v_a h_a - v_a^2 d_a = (u_a - v_a) d_a + 2u_a v_a h_a$. The contribution of A-a to the variance of the population will thus be

$$u_a^2 d_a^2 + 2u_a v_a h_a^2 + v_a^2 d_a^2 - [(u_a - v_a) d_a + 2u_a v_a h_a]^2$$

which reduces to $\quad 2u_a v_a [d_a^2 + 2(v_a - u_a) d_a h_a + (1 - 2u_a v_a) h_a^2]$

which in its turn can be rewritten as

$$\begin{aligned} &2u_a v_a [d_a^2 + 2(v_a - u_a) d_a h_a + (v_a - u_a) h_a^2 + 2u_a v_a h_a^2] \\ &= 2u_a v_a [d_a + (v_a - u_a) h_a]^2 + 4u_a^2 v_a^2 h_a^2. \end{aligned}$$

Where the genes are independent in their action and uncorrelated in their distribution within the population, the total heritable variation will be the sum of a series of such terms, one from each gene pair, namely

$$V_R = \text{S}\, 2uv\, [d + (v - u)h]^2 + \text{S}\, 4u^2 v^2 h^2.$$

If we now put $D_R = \text{S}\, 4uv\, [d + (v - u)h]^2$ and $H_R = \text{S}\, 16u^2 v^2 h^2$ the heritable variance becomes $\frac{1}{2}D_R + \frac{1}{4}H_R$, and apart from sampling variations this heritable variance will be constant from one generation to another. We have already met these expressions for D_R and H_R earlier, when we were discussing undefined diallels in Section 17.

Where $u = v = \frac{1}{2}$ for all genes, as in the F_2 of a cross between two true-breeding parental lines, these expressions for D_R and H_R reduce to $\text{S}(d^2)$ and $\text{S}(h^2)$ and the heritable variance itself becomes $\frac{1}{2}D + \frac{1}{4}H$ as already found for V_{1F2}. This is indeed as it should be since an F_2 can be regarded as the special case of a population where necessarily $u = v = \frac{1}{2}$. It will thus be seen that if and only if $u = v$ the contributions made to the heritable variance by d and h will be capable of complete separation. Where $u \neq v$ D_R will always be affected by h, and H_R will be correspondingly less than the sum of h^2. D_R will be greater than $S(d^2)$ where $S(v-u)h$ is positive which will happen when the dominant genes are in general rarer than their recessive alleles. Equally D_R will be less than $D = S(d^2)$ where $S(v-u)h$ is negative, that is where the dominant genes are in general commoner than their recessive alleles, (Fig. 20). In fact if in general $h > d$, D_R will become 0 where $u = (d+h)/2h$.

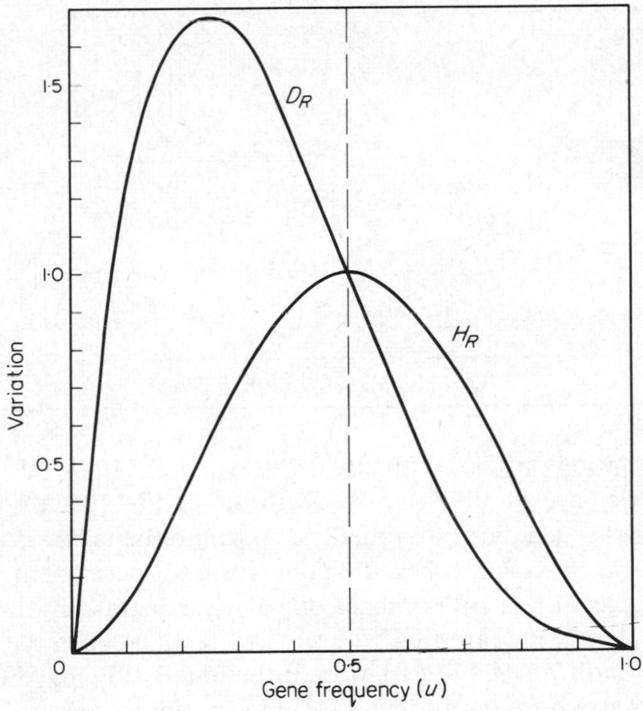

Fig. 20. Change in the contribution made by a gene pair to D_R and H_R according to u, the frequency of the dominant allele, in a randomly breeding population, where $d = h = 1$.

The value of D_R thus depends not only on the effects of the various genes of the system when in the homozygous state, which we denote by d, but also on h, their effects when heterozygous, and on the allele frequencies u and v. Only when either $h = 0$ or $u = v = \frac{1}{2}$ (or of course when both conditions are satisfied) does $D_R = D = S(d^2)$. Thus D_R is not in general the additive variation as we have defined and used this term in the earlier chapters.

It is nevertheless frequently referred to as such. As so used it is the

TABLE 51.

The pair matings in a randomly breeding population in respect of a single gene difference

		Female parents			
		AA	Aa	aa	
	Frequency	u^2	$2uv$	v^2	
Male parents	AA u^2	u^4 d 0	$2u^3v$ $\frac{1}{2}(d+h)$ $\frac{1}{4}(d-h)^2$	u^2v^2 h 0	Frequency mean variance
	Aa $2uv$	$2u^3v$ $\frac{1}{2}(d+h)$ $\frac{1}{4}(d-h)^2$	$4u^2v^2$ $\frac{1}{2}h$ $\frac{1}{2}d^2+\frac{1}{4}h^2$	$2uv^3$ $\frac{1}{2}(h-d)$ $\frac{1}{4}(d+h)^2$	
	aa v^2	u^2v^2 h 0	$2uv^3$ $\frac{1}{2}(h-d)$ $\frac{1}{4}(d+h)^2$	v^4 $-d$ 0	
		Overall mean $(u-v)d+2uvh$			

additive variation only in a statistical sense, and not in the genetical sense that we have adopted. Unlike D it is not a direct measure of the variation that is genetically fixable and so cannot be taken as a certain guide to the innate capacity of the population for permanent genetical change by selection or other means of genetical manipulation.

If a population is composed of a series of families, each the progeny of a pair of individuals, the variation of the population may be subdivided into variation within families and variation between them. Where mates come together at random in relation of their genotypes, there are nine possible types of mating in respect of a single gene difference (Table 51). In respect of the parental and progeny genotypes, these nine types of mating fall into six classes which are recognizable as equivalent to the two parental F_1, F_2 and two back-cross families; whose means and variances are already known from Chapter 3.

The mean variance within these families is found directly from Table 51 by summing the products of the frequencies of the families and their variances, to give

$$u_a^4.0 + 4u_a^3v_a.\tfrac{1}{4}(d_a-h_a)^2 + 2u_a^2v_a^2.0 + 4u_a^2v_a^2.(\tfrac{1}{2}d_a^2+\tfrac{1}{4}h_a^2) + 4u_av_a^3.\tfrac{1}{4}(d_a+h_a)^2 + v_a^4.0$$

which reduces to

$$u_a v_a d_a^2 + u_a v_a (v_a - u_a) d_a h_a + (u_a v_a - u_a^2 v_a^2) h_a^2$$
$$= u_a v_a [d_a + (v_a - u_a) h_a]^2 + 3 u_a^2 v_a^2 h_a^2.$$

On summing over all relevant genes this yields $\frac{1}{4}D_R + \frac{3}{16}H_R$.

The variance of family means, measuring the variation between families is similarly found from the frequencies and means of the types of family, as

$$u_a^4 d_a^2 + 4u_a^3 v_a (\tfrac{1}{2}d_a + \tfrac{1}{2}h_a)^2 + 2u_a^2 v_a^2 h_a^2 + 4u_a^2 v_a^2 (\tfrac{1}{2}h_a)^2$$
$$+ 4u_a v_a^3 (-\tfrac{1}{2}d_a + \tfrac{1}{2}h_a)^2 + v_a^4 d_a^2 - [(u_a - v_a) d_a + 2u_a v_a h_a]^2$$

the last term being the correction for the overall mean. This reduces to

$$u_a v_a [d_a + (v_a - u_a) h_a]^2 + u_a^2 v_a^2 h_a^2.$$

Summing over all relevant genes we obtain $\frac{1}{4}D_R + \frac{1}{16}H_R$.

These two variances sum to give $\frac{1}{2}D_R + \frac{1}{4}H_R$, the total heritable variance of the population, as obviously they must. In the special case of $u = v = \frac{1}{2}$, where D_R becomes D and H_R becomes H, the two variances become respectively $\frac{1}{4}D + \frac{3}{16}H$ and $\frac{1}{4}D + \frac{1}{16}H$ which we have already found for V_{2S3} and V_{1S3}. Thus such families within a population may be regarded as the general case of biparental families obtained from an F_2, just as the population itself corresponds to the general case of the F_2.

The members of a single family are distinguishable in the population as full-sibs. The covariance of such full-sibs may be obtained directly, but it is simpler to note that where a population is divided into groups of like status, such as our families of full-sibs, the mean covariance of two members of the same group can be shown to equal the variance of the group means. We can therefore immediately write down the covariance of full-sibs taken over the population as a whole as $\frac{1}{4}D_R + \frac{1}{16}H_R$.

Where the mating system of a population is such that any parent may leave a number of offspring, the second parent of which is however prospectively different for each of them, this second parent being drawn at random from the population, full-sibs will be rare but groups with one common parent, and composed therefore of what are termed half-sibs, may be recognized. In such a case the second parents may be regarded as providing a set of gametes having the population frequencies of A and a, namely u_a and v_a. The properties of these families will thus be as shown in Table 52. The contributions of A-a to the mean variance of the single parent families and to the variance of their means are given at the foot of

TABLE 52.

Families of individuals having one parent in common, and thus composed of half-sibs (HS), in a randomly breeding population in respect of a single gene difference

[Note: since the second parents of the progeny of any common parent are drawn at random from the population they may be regarded as providing an array of uA + va gametes for fusion with those of the common parent]

	Common parent			
Progeny	AA u^2 d	Aa $2uv$ h	aa v^2 $-d$	Frequency in population Phenotype
AA	u d	$\frac{1}{2}u$ d	0 —	Frequency in family Phenotype
Aa	v h	$\frac{1}{2}(u+v)$ h	u h	
aa	0 —	$\frac{1}{2}v$ $-d$	v $-d$	
Family mean	$ud + vh$	$\frac{1}{2}[(u-v)d + h]$	$uh - vd$	
Family variance	$uv(d-h)^2$	$2uvd^2$ $+\frac{1}{4}[(v-u)d+h]^2$	$uv(d+h)^2$	

Mean variance $= \frac{3}{2}uv[d+(v-u)h]^2 + 4u^2v^2h^2 \rightarrow V_{2HSR} = \frac{3}{8}D_R + \frac{1}{4}H_R$

Variance of means (= covariance of half-sibs) $= \frac{1}{2}uv[d+(v-u)h]^2 \rightarrow V_{1HSR} = W_{HSR} = \frac{1}{8}D_R$

Covariance of parent and offspring $= uv[d+(v-u)h]^2 \rightarrow W_{POR} = \frac{1}{4}D_R$

the table, and it will be seen that on summing over the relevant gene differences, the heritable portion of mean variance becomes $\frac{3}{8}D_R + \frac{1}{4}H_R$ and that of the variance of the family means becomes $\frac{1}{8}D_R$. These two variances sum to $\frac{1}{2}D_R + \frac{1}{4}H_R$ the heritable variance of the population, as indeed they clearly should. The covariance of the half-sibs, of which these families are composed, will of course be the same as the variance of the family means, namely $\frac{1}{8}D_R$.

One further statistic may be found from Table 52. The covariance of a single parent and its offspring is found as the covariance of the common parent and the mean of its offspring as set out in the table. This is clearly

$$u_a^2 d_a [u_a d_a + v_a h_a] + u_a v_a h_a [(u_a - v_a) d_a + h_a] + v_a^2 d_a [v_a d_a - u_a h_a] - [(u_a - v_a) d_a + 2u_a v_a h_a]^2 = u_a v_a [d_a + (v_a - u_a) h_a]^2$$

the correction term being the square of the population mean, since this is the mean of all the parents as well as the mean of all their progeny. Summing over all the relevant gene differences then shows the covariance of parent and offspring to be $\frac{1}{4}D_R$.

All that remains to complete these formulations of different variances and covariances derivable from the population is to add in the appropriate items for non-heritable and sampling variation. Here as in our earlier consideration (Section 12) we must distinguish between the non-heritable variation among members of the same family and that between families. If we denote by E_w, the non-heritable variance within families, the mean variances of full-sib and half sib-families become respectively $\frac{1}{4}D_R + \frac{3}{16}H_R + E_w$ and $\frac{3}{8}D_R + \frac{3}{16}H_R + E_w$. Where E_b is the non-heritable variance between families the variances of family means must obviously include E_b. They will, however, also include an item for sampling variation which will of course be $\frac{1}{n}\overline{V}$, where $\overline{V}$ denotes the relevant mean variance and n is the number of individuals in the family or the harmonic mean of these numbers if they vary from one family to another. Thus if V_{1SR} and V_{2SR} stand for the variance of the mean and the mean variance of full-sib(S) families as observed in a randomly breeding population

$$V_{1SR} = \tfrac{1}{4}D_R + \tfrac{1}{16}H_R + E_b + \tfrac{1}{n}(V_{2SR})$$

and

$$V_{2SR} = \tfrac{1}{4}D_R + \tfrac{3}{16}H_R + E_w.$$

Similarly for half-sib families, denoted by the inclusion in the suffix of HS in place of S standing for full-sibs,

$$V_{1HSR} = \tfrac{1}{8}D_R + E_b + \tfrac{1}{n}(V_{2HSR})$$

and

$$V_{2HSR} = \tfrac{3}{8}D_R + \tfrac{1}{4}H_R + E_w.$$

Now if the individuals from a family are distributed independently of one another across the range of the environments throughout their lives there will be no cause of non-heritable variation between families additional to those within, and $E_b = 0$. But if families remain together, perhaps also enjoying parental attention as in many animal species, or being endowed by the mother with nutritional resources on which to draw during early life as happens in both plants and animals, there will be non-heritable differences between families which go beyond those within: E_b is then > 0 and will be reflected by a corresponding increase in the variance of family means. Furthermore, since members of the same family will share the same environment in respect of such family effects while members of different families will not, their covariance will reflect E_b also, whether they are sibs or half-sibs. An E_b component must therefore also be included in these covariances which thus become

$$W_{SR} = \tfrac{1}{4}D_R + \tfrac{1}{16}H_R + E_b$$

and

$$W_{HSR} = \tfrac{1}{8}D_R + E_b.$$

And if offspring in some measure share the environment of their parents the same will be true to a corresponding extent of the parent/offspring covariance, which must thus be written as

$$W_{POR} = \tfrac{1}{4}D_R + E_b'$$

the prime indicating that the non-heritable effects common to parent and offspring may not be just the same as that shown by members of the same progeny.

These various results are collected together in Table 53. Two points remain to be made about them. First, where nutritional resources for early life are provided by the mother, or where parental attention is provided and it is not the same from mother and father, the E_b component in the covariance of half-sibs will be different according to whether the common parent is mother or father. Secondly the non-heritable variance of the population as a whole will be $E_w + E_b$ since each individual will reflect both effects. Where $E_b = 0$ this non-heritable component of

TABLE 53.

Composition of variances and covariances in a randomly breeding population

Relationship	Statistic
Full-sib families (both parents common)	$V_{1SR} \doteq \tfrac{1}{4}D_R + \tfrac{1}{16}H_R + E_b + \tfrac{1}{n} V_{2SR}$
	$V_{2SR} = \tfrac{1}{4}D_R + \tfrac{3}{16}H_R + E_w$
	$W_{SR} = \tfrac{1}{4}D_R + \tfrac{1}{16}H_R + E_b$
Half-sib families (one parent common)	$V_{1HSR} = \tfrac{1}{8}D_R + E_b + \tfrac{1}{n} V_{2HSR}$
	$V_{2HSR} = \tfrac{3}{8}D_R + \tfrac{1}{4}H_R + E_w$
	$W_{HSR} = \tfrac{1}{8}D_R + E_b$
Parent and offspring	$W_{POR} = \tfrac{1}{4}D_R + E_b'$
Whole population	$V_R = \tfrac{1}{2}D_R + \tfrac{1}{4}H_R + E_w + E_b$

V_R will of course reduce to E_W which will obviously cover all the environmental differences among the individuals of the population.

31. Human populations

In most species we can carry out the analysis of a population by experimental means, that is by using families obtained from controlled matings and by adopting experimental designs that will enable us to disentangle the various heritable components of variation both from one another and from the non-heritable components. How this can be done will be seen in a later section, but first we must look at a species, our own, where neither controlled mating nor the controlled distribution of individuals or groups of individuals among the differing environments is possible. Despite these limitations, man offers many advantages for the study of populations. In particular, in our present context, we know more about the variation to be observed in human populations than in those of any other species; we can trace a more complex and wider range of relationships than in any other species; we can observe human mates even if we cannot control their choices, and so can detect and measure departures from random mating among them; and we can detect with some confidence monozygotic twins and distinguish them from their dizygotic counterparts.

The classical approach to the genetical analysis of continuous variation in man is by the use of correlations between individuals of known relationships, an approach that was initiated by Galton a hundred years ago and put to such good use by Fisher in 1918. In principle, such correlations are obtainable for many different degrees of relationship, but in practice relatively few have been used. We will illustrate this approach and its limitations using two genetical relationships, those between parent and offspring and between full-sibs. (To these we will add the correlation between spouses which is zero when mating is at random but which is commonly observed to depart from this expectation.) Fisher (1918) records that Pearson and Lee observed the correlation between parent and offspring (r_{po}) to be 0.4180, and that between full-sibs (r_{ss}) to be 0.4619 in respect of the cubit measurement, that is the length of the forearm from elbow to fingertip. If we assume mating to be at random we have from Table 53

$$r_{po} = \frac{W_{POR}}{V_R} = \frac{\frac{1}{4}D_R + E_b'}{\frac{1}{2}D_R + \frac{1}{4}H_R + E_w + E_b} = 0.4180$$

and $$r_{ss} = \frac{W_{SR}}{V_R} = \frac{\frac{1}{4}D_R + \frac{1}{16}H_R + E_b}{\frac{1}{2}D_R + \frac{1}{4}H_R + E_w + E_b} = 0.4619.$$

The denominator used in finding r_{po} is, of course, the geometric mean of the variances of parents and offspring, but when single parents and single offspring are used in finding W_{POR} and these are a fair sample from the population, the variances of both parents and offspring, and hence their geometric mean, will all be V_R as shown. The same argument applies to the denominator used in finding r_{ss}.

If we could further assume that the non-heritable variation between individuals from different families was no greater than that between individuals from the same family, i.e. $E_b = E_b' = 0$ these equations would reduce to

$$r_{po} = \frac{\frac{1}{4}D_R}{\frac{1}{2}D_R + \frac{1}{4}H_R + E_w} = 0.4180$$

and $$r_{ss} = \frac{\frac{1}{4}D_R + \frac{1}{16}H_R}{\frac{1}{2}D_R + \frac{1}{4}H_R + E_w} = 0.4619.$$

Although we would still have three parameters with only two equations and so be unable to estimate the numerical values of the parameters, we could find their values relative to one another or more usefully find the relative contributions that D_R, H_R and E_w made to V_R the total variance of the population. Thus

$$\tfrac{1}{4}D_R = r_{po}V_R = 0.4180\ V_R$$

$$\tfrac{1}{16}H_R = (r_{ss} - r_{po})\ V_R = 0.0439\ V_R.$$

Then $\frac{1}{2}D_R$ would be 0.8360 V_R and $\frac{1}{4}H_R$ would be 0.1756 V_R leaving $E_w = -0.0116\ V_R$, and we should conclude that the variation in the population was almost entirely heritable, with E_w very small and our estimate of it becoming negative through sampling variation.

The assumption that non-heritable variation between individuals from different families is no greater than that between individuals from the same family would, however, be very difficult to sustain in man: indeed our experience would point strongly the opposite way. We cannot therefore set $E_b = E_b' = 0$ and must use the full equations which include these parameters. When we do so we find that in place of D_R, H_R and E_w, as they appear in the solutions of the equations when simplified by the omission of E_b and E_b', we have $D_R + 4E_b'$, $H_R + 16(E_b - E_b')$ and $E_w - 3E_b + 2E_b'$, giving as the partition of V_R

$$\tfrac{1}{2}D_R + 2E_b' = 0.8360\,V_R$$
$$\tfrac{1}{4}H_R + 4(E_b - E_b') = 0.1756\,V_R$$
$$E_w - 3E_b + 2E_b' = -0.0116\,V_R.$$

Furthermore it is impossible to take the analysis further because it is impossible to separate the estimate of D_R from E_b', that of H_R from $E_b - E_b'$ and that of E_w from E_b and E_b' although we might note that the estimate of H_R is less affected by non-heritable differences between families than is that of D_R, and indeed is completely free of non-heritable effects if $E_b' = E_b$. Thus our conclusions must be revised: the results show that the variation of the population is almost entirely accounted for not just by genetic differences but by genetic differences plus the non-heritable differences between families, and until we can find some means of separating the genetical effects from the non-genetic differences between families we can take the analysis no further. How this can be done using twin studies in particular we shall see in a moment, but before proceeding to this we must look at a further complication in the analysis of correlations among human relatives.

In deriving our formulae for r_{po} and r_{ss} we assumed mating to be at random. We know, however, that this assumption is not fully justifiable. In respect of the cubit measurement, Pearson and Lee found that there is a correlation between spouses of $r_{FM} = 0.1977$. In other words there is positive assortatitve mating in respect of this character: there is a tendency for like to mate with like in respect of this (as indeed of most other) characters in man. The effects of assortative mating may be complex and a detailed consideration of them is beyond the scope of our present consideration, but two results may be noted. First, where a large number of gene differences are involved in the variation, assortative mating does not alter the contribution of dominance deviations (h's) to the variation. Nor does it affect the additive variation within families. It does however, change the additive variation between families, raising it with positive assortative mating, i.e. when r_{FM} is positive, and lowering it with negative assortative mating, i.e. when r_{FM} is negative. Secondly, since both the r_{po} and r_{ss} depend on comparisons between families, they will be increased by positive assortative mating such as has been observed for the cubit measurement. So any analysis of variation based on the assumption of random mating, such as the one we have carried out, will overestimate the additive genetic component if positive assortative mating is in fact in operation. We can obtain an idea of the extent of this overestimation by noting that, other things being equal, r_{po} is proportional to $(1 + r_{FM})$. Since for the cubic measurement r_{FM} was found to be 0.1977, r_{po} will be raised to approximately (1 + 0.1977) times the value it would have shown had the population been truly random mating as we assumed. Thus the true contribution of $D_R + 4E_b'$ to the population variance should have been about 1/1.1977 or say 5/6 the value we found, i.e. 0.70 instead of 0.84 as we calculated it on the assumption of random mating. How this change should be apportioned between D_R and E_b' cannot of course be determined, since we cannot separate these parameters in the analyses.

32. The use of twins

If twins arise at random in the population the total variation among a sample of them, assuming random mating, will be the same as in the population as a whole, namely, $\frac{1}{2}D_R + \frac{1}{4}H_R + E_w + E_b$. Where we have monozygotic (identical) twins raised together (MZT) in their natural family groupings the variation within the pair is entirely due to non-heritable causes operating within a family and will therefore on average, be E_w. All the remaining variation, that is $\frac{1}{2}D_R + \frac{1}{4}H_R + E_b$, will be between the means of families of twins. This expectation, however, like all theoretical expectations assumes very large family sizes whereas with families of twins the family size is very small, indeed it is always two. The expected variance of family means must, therefore, have added to it half the mean variance within families, that is $\frac{1}{2}E_w$. (For families of size n this would be $\frac{1}{n}E_w$ and, of course, where n is very large this reduces for all practical purposes to zero.)

If we now have monzygotic twins that have been raised apart (MZA) and these are a random sample of all twins, the difference within twin pairs will still be entirely non-heritable but it will now include both within and between family components, that is both E_w and E_b. Providing that the separated twins are distributed at random among families in the population the mean variance within twin pairs will be $E_w + E_b$ and the variation between pairs of twins means after correcting for the effect of families of size two, will therefore be

$$\tfrac{1}{2}D_R + \tfrac{1}{4}H_R + \tfrac{1}{2}(E_w + E_b).$$

With monozygotic twins raised apart we can separate the environmental and the heritable sources of variation independently of the model assumed for the latter and therefore independently of the kind of gene action and interaction present and the mating system. Even if we assume, however, that an additive-dominance model with random mating is the appropriate model giving $\frac{1}{2}(E_w + E_b)$ and $\frac{1}{2}D_R + \frac{1}{4}H_R$ respectively as the non-heritable and heritable components this does not enable us to separate the additive and dominance variation. And even if we further combine data from MZT and MZA we still cannot separate D_R and H_R although we may now separate E_w from E_b.

Shields (1962) reports a measure of Neuroticism in man for 29 pairs of monozygotic female twins raised together, 26 pairs raised apart and 14 pairs of male twins raised apart which have been analysed by Jinks and Fulker (1970). The mean variances within families $\bar{V}_F$ and the variances of family means $V_{\bar{F}}$ are as follows:

		Females	Males	Expectations
MZT	$V_{\bar{F}}$	11.0819	–	$\frac{1}{2}D_R + \frac{1}{4}H_R + \frac{1}{2}E_w + E_b$
	$\bar{V}_F$	8.1207	–	E_w
MZA	$V_{\bar{F}}$	14.5608	14.7307	$\frac{1}{2}D_R + \frac{1}{4}H_R + \frac{1}{2}E_w + \frac{1}{2}E_b$
	$\bar{V}_F$	9.6635	5.0000	$E_w + E_b$.

The estimates of the total variance of the MZT and the two MZA samples are not significantly different. This is expected on the model since they should all be estimates of $\frac{1}{2}D_R + \frac{1}{4}H_R + E_w + E_b$. Equally the mean scores in the three samples do not differ significantly, being 9.72, 11.86 and 10.71 respectively. This too is expected on our model which assumes that all three samples are drawn from the same population and, therefore, have the same genetical and environmental sources of variation. This does not of course mean that the specification of the genetical component as $\frac{1}{2}D_R + \frac{1}{4}H_R$ and the environmental component as $E_w + E_b$ is necessarily adequate but that the genetical and environmental components are the same for all three samples whatever their compositions. We can, therefore, regard the males and females as replicate estimates of the statistics for the purposes of analysis. For twins raised apart therefore

$$V_{\bar{F}} = \tfrac{1}{2}D_R + \tfrac{1}{4}H_R + \tfrac{1}{2}E_w + \tfrac{1}{2}E_b = 14.6458$$

and

$$\bar{V}_F = E_w + E_b = 7.3318$$

hence

$$\tfrac{1}{2}D_R + \tfrac{1}{4}H_R = 10.9799.$$

Some 60% of the variation is, therefore, due to heritable differences and 40% due to environmental differences. Since we have not needed to take account of the make up of either the heritable or environmental portions in arriving at this partition the result would be the same irrespective of the model assumed for either.

This still remains true for the genetical component even on combining the twins raised together and apart. On our simple model we now have four statistics but only three parameters since $\frac{1}{2}D_R$ and $\frac{1}{4}H_R$ are still inseparable. We can therefore, obtain least squares estimates by the normal procedures (Section 9). These are

$$\begin{aligned} \tfrac{1}{2}D_R + \tfrac{1}{4}H_R &= 10.0291 \\ E_w &= 8.7546 \\ E_b &= -2.0568. \end{aligned}$$

We can now calculate the expected values of the four statistics and we have one degree of freedom for comparing the observed and expected values. From replicate statistics (males and females for MZA) we have an error variance for two degrees of freedom against which to test the significance of the discrepancy between observed and expected values.

		Observed	Expected	Deviation
MZT	$V_{\bar{F}}$	11.0819	12.3496	−1.2677
	$\bar{V}_F$	8.1207	8.7546	−0.6339
MZA	$V_{\bar{F}}$	14.6458	13.3781	1.2677
	$\bar{V}_F$	7.3318	6.6978	0.6339

The SS of deviations is $(-0.6339)^2 + (-1.2677)^2 + (0.6339)^2 + (1.2677)^2$

$$= 4.0152 \text{ for 1 df}$$

and the SS for replicates is $\frac{1}{2}(14.5608 - 14.7307)^2 + \frac{1}{2}(9.6635 - 5.0000)^2$

$$= 10.8871 \text{ for 2 df.}$$

However, for two of the four statistics to which we are fitting the model we are working with the averages of two replicates and hence the replicate mean square appropriate for testing the deviation mean square is

$$\tfrac{1}{2}\left(\frac{10.8871}{2} + \frac{10.8871}{4}\right) = 4.0827.$$

The two mean squares clearly do not differ.

The deviations, therefore, are no greater than would be expected to arise from error variation and we can conclude that the model fits adequately. The deviation mean square and the error mean square being homogeneous may be pooled to give an error variance of 4.0602 for 3 df. By multiplying this by the appropriate coefficients on the leading diagonal of the inverted matrix (see Section 9) we obtain the error variance of each of the estimates of the three components and hence their standard errors. These are:

$$\tfrac{1}{2}D_R + \tfrac{1}{4}H_R = 10.0291 \pm 2.0400 \quad t_{(3)} = 4.92 \quad P = 0.01 - 0.02$$
$$E_w = 8.7546 \pm 1.9116 \quad t_{(3)} = 4.58 \quad P = 0.02$$
$$E_b = -2.0568 \pm 2.5488 \quad t_{(3)} = 0.81 \quad P = 0.40 - 0.50.$$

Thus, although the error variance is based on very few degrees of freedom we can see that the estimate of the genetic component and E_w are significant while the negative E_b is not. These estimates are in good agreement with those obtained from our earlier analysis of MZA's alone.

Before proceeding further we should reiterate that while we make the usual assumptions of random mating and no non-allelic interactions to arrive at the expectations for the heritable variation, the process of fitting the model and the estimates obtained would be unchanged irrespective of the mating system and the nature of the gene action or interaction. As we pointed out earlier this partitioning of the variation into a heritable and a non-heritable component makes no assumptions about the nature of either.

Examination of the deviations of observed and expected values of the four statistics shows that they are identical in value but opposite in sign between the statistics from twins raised together and twins raised apart. The 1 df for testing the significance of these deviations is in fact testing the difference between the total variance components of the two types of twins which we expect to be identical. This test is, therefore, equivalent to our earlier test of the homogeneity of the three total variances and not surprisingly they agree in finding the model adequate.

We could improve these estimates and increase the power of the test of significance by repeating the estimations using a weighted least squares procedure (Section 9). With such a good fit to the model as is shown by these data, however, the improvement can only be marginal. It is more important to consider an assumption implicit in the model whose validity has not been tested by the test of goodness of fit of the model. This is the assumption of no genotype × environment interaction. Our tests of the goodness of fit of the model are in effect tests of the homogeneity of the total variances. Since the total variances are expected to have the same genetical and environmental components irrespective of the constitution of these components they are also expected to have the same genotype × environment interaction components. We cannot, therefore, test the assumption of no genotype × environment interactions by testing the homogeneity of the total variances, and we have therefore so far no test of this assumption. Nevertheless, it is possible to provide a sensitive test for certain kinds of genotype × environment interactions.

The difference between a pair of monzygotic twins is solely environmental in origin. In the absence of genotype × environment interactions, therefore, the magnitude of this difference should be independent of the genotypes of the twin pairs. This expectation is identical with the expectation that in the absence of genotype × environmental interactions the variation between the individuals of a family should be the same for all pure breeding lines and the F_1's produced by crosses between them (Chapter 6). Our measure of the genotypic differences between twins is the difference between family means of twins raised apart. We can test the assumption of no genotype × environment interaction, therefore, by testing for the independence of the means (or sums) and differences of twin pairs where the twins have been raised apart (Jinks and Fulker, 1970). Since in general we cannot cross classify twins over families, the signs we allocate the differences are arbitrary. We can make them all positive by always taking the smaller twin score from the larger or all negative by doing the reverse. Equally, we can take them at random in which case approximately half will be positive and half negative. We shall adopt the convention of making them all positive.

We can examine the sums and differences for twin pairs for evidence of non-independence by plotting one against the other for all twin pairs. Non-independence would then show itself by the points departing from a random scatter by being distributed along a line or curve. Statistically we can detect non-independence by calculating the correlation between sums and differences over the twin pairs. For Neuroticism this leads to a correlation of $r = 0.0583$ over the 40 pairs of MZA for 38 df. Clearly there is no relationship and hence the magnitude of the environmentally caused differences is independent of the genetical differences. That is, there is no evidence of genotype × environment interaction.

Monozygotic twins raised together provide a similar but less comprehensive test of the assumption because the differences include only within family environmental effects (E_w), and the sums include the common environmental effects that arise from sharing the same family environment (E_b). Nevertheless, if the scatter diagram and correlation show no relationship we can still conclude that genotype × environment interactions are absent. If, however, they reveal a relationship we cannot claim that the presence of genotype × environment interactions has been unambiguously demonstrated because, no matter how unlikely this may be, such a relationship could have arisen because of the non-independence of the within and between family environmental components. For the Neuroticism data the MZT's confirm the absence of genotype × environment interactions ($r = 0.1489$).

We can, therefore, claim to have separated the genetical and environmental components of variation for Neuroticism without making any untestable assumptions.

We can extend the analysis indefinitely by adding samples from other kinds of families and other kinds of relationships. In particular, in the present context, we can extend it to include dizygotic twins. The statistics obtainable from dizygotic twins have the same expectations as those of full-sibs for the standard within and between family variances provided that dizygotic twins arise at random in the population. By adding these statistics to our earlier analysis of monozygotic twins we can now test the adequacy of a model that assumes random mating and an additive-dominance model of the gene action.

Shields (1962) gives the Neuroticism scores of 16 pairs of female dizygotic twins raised together (DZT) and Jinks and Fulker have presented a combined analysis of these with the data from the monozygotic twins. The mean score and the total variance do not differ from those of the MZT and MZA and all three can, therefore, be regarded as random samples drawn from the same population. The six observed statistics, after pooling males and females as before, and their expectations on the model are

Source		Observed	Expected when $H_R = 0, E_b = 0$	Model
MZT	$V_{\bar{F}}$	11.0819	13.0290	$\frac{1}{2}D_R + \frac{1}{4}H_R + \frac{1}{2}E_w + E_b$
	$\bar{V}_F$	8.1207	7.7199	E_w
MZA	$V_{\bar{F}}$	14.6458	13.0290	$\frac{1}{2}D_R + \frac{1}{4}H_R + \frac{1}{2}E_w + \frac{1}{2}E_b$
	$\bar{V}_F$	7.3317	7.7199	$E_w + E_b$
DZT	$V_{\bar{F}}$	11.7828	10.7368	$\frac{3}{8}D_R + \frac{5}{32}H_R + \frac{1}{2}E_w + E_b$
	$\bar{V}_F$	13.8552	12.3045	$\frac{1}{4}D_R + \frac{3}{16}H_R + E_w$.

Fitting the full model by least squares procedures confirms our earlier conclusion that E_b is not significantly different from zero and also reveals that H_R is not significant. A D_R, E_w model may, therefore, be fitted which with six observed statistics leaves 4 df for testing the adequacy of the model against the replicate error. The least squares estimates are

$$D_R = 18.6248 \quad \text{and} \quad E_w = 8.1605.$$

We can of course go further and obtain improved estimates of D_R and E_w by weighting the observed statistics by their amounts of information (Section 9). This method gives

$$D_R = 18.3380 \pm 4.9884 \quad c = 3.68 \quad P < 0.001$$
$$E_w = 7.7199 \pm 1.2755 \quad c = 6.05 \quad P < 0.001$$

which agrees with the estimates from the simpler calculations just given and with the estimates based on monozygotic twins alone. The test of the fit of the model based on the comparison of the observed and expected statistics from the weighted estimation leads to an approximate $\chi^2_{[4]} = 1.3717$ (P = 0.80) which confirms once more the adequacy of the simple model.

These four degrees of freedom for testing the adequacy of the model are made up of two parts. Two degrees of freedom are testing the effect of omitting H_R and E_b from the full model and two are testing the equality of the total variance components of the three types of twins which are expected to be equal on the model. Since the D_R, E_w model is adequate this confirms that H_R and E_b are not significantly different from zero and that the total variance components do not differ significantly. This in turn confirms the earlier test of the homogeneity of the three total variances. We can conclude, therefore, that dominance and the family environment have no detectable effects on the Neuroticism score and that all three types of twins, that is MZT, MZA and DZT, are subject to the same heritable and environmental sources of variation. Hence, the results provide no evidence for the often assumed greater environmental heterogeneity experienced by dizygotic relative to monozygotic twins.

We have now considered three sets of data each of which allow us to separate heritable from non-heritable sources of variation. In each set, however, it is the presence of monozygotic twins raised apart that has permitted this partitioning. Indeed, as we have seen, we can make this partitioning solely on the basis of MZA scores and at the same time have available the best test for genotype × environment interactions. What we cannot do however, without involving other types of twin data or other kinds of family relationships is to test any other assumptions we may care to make about the sources of variation, mating system, etc.

Providing that we retain MZT and DZT scores we can substitute dizygotic twins reared apart (or full-sibs reared apart) DZA for MZA to obtain an almost equally effective test of the assumptions and estimates of the parameters of the additive-dominance model of gene action if adequate. The expectations of the two variances for DZA on this model for a randomly mating population are

$$V_{\bar{F}} = \tfrac{3}{8}D_R + \tfrac{5}{32}H_R + \tfrac{1}{2}E_w + \tfrac{1}{2}E_b$$
$$\bar{V}_F = \tfrac{1}{4}D_R + \tfrac{3}{16}H_R + E_w + E_b.$$

As we have seen twins, or alternatively full-sibs, raised apart are invaluable for unambiguously separating heritable and non-heritable sources of variation. The extent to which they allow us to achieve this objective, however, rests on the validity of the assumption that the two individuals of each twin pair are distributed at random among the family environments present in the population. We can test whether 'foster' homes are a random sample of family environments by comparing their mean and variance for any particular measure with those of a random sample of 'own' homes. There are a variety of measures we can use for this purpose. We could, for example, measure the physical environment directly using an index such as socio-economic class that has been developed by social scientists for comparing family environments. Equally, of course, we could measure the environment biologically as we did in Chapter 6 to analyse genotype × environment interactions. One measure might then be the phenotypes of the parents, either biological or foster, who provide the home environment in respect of the character in question.

This can tell us whether foster homes are a random sample. It does not, however, tell us whether the separated twins were allocated to this sample of foster homes at random. That is, whether there is a 'placement' effect because successful attempts have been made to match the fostered individuals with the foster home. In such a case the separated twins would have been raised independently but in similar family environments. In order to test for such effects we would have to look for a correlation between the family environments of separated twins. Our measure of the family environments would again be based on an environment index or the phenotypes of the foster parents. Only if the correlations were non-significant could we conclude that the separated twins provided a valid estimate of the total environmental effects.

Much of the available twin data consists of MZT and DZT and as we have already noted an unambiguous analysis of such data is not generally possible because the simplest additive-dominance, random mating model has four parameters and we can fit only three as a maximum. If one of the two parameter models fits, for example E_w and E_b or E_w and D_R, and the others fail we can be confident of the results. If, however, all the two parameter models fit equally well or fail equally badly no unambiguous conclusion is possible. We have no basis for choosing between the alternative two parameter models and all three parameter models are equally satisfactory since all would lead to perfect fit solutions. What can and cannot be achieved in these circumstances is well illustrated by the work of N. G. Martin (1975).

Even more typical of the kind of twin data found in the literature are the observations of Holt (1952) on the number of palm print ridges in man which are presented in the form of correlations for MZT, DZT and full-sib families. Although correlations provide a useful summary of the data and are widely used in human genetics, they are not a good starting point for an analysis. In particular we cannot carry out any of the tests of assumptions that depend on a comparison of total variances. As correlations the data have been standardized to the same unit total variance for all kinds of families and at the same time we lose one statistic from each kind of family.

For this character mating is known to be at random. The correlation for monozygotic twins on the additive-dominance model is therefore

$$r = 0.96 = \frac{\frac{1}{2}D_R + \frac{1}{4}H_R + E_b}{\frac{1}{2}D_R + \frac{1}{4}H_R + E_w + E_b} \quad \text{which can be rewritten}$$

$$\tfrac{1}{2}D_R + \tfrac{1}{4}H_R + E_b = 0.96(\tfrac{1}{2}D_R + \tfrac{1}{4}H_R + E_w + E_b).$$

Similarly, from Holt's correlations,

$$\tfrac{1}{4}D_R + \tfrac{1}{16}H_R + E_b = 0.47(\tfrac{1}{2}D_R + \tfrac{1}{4}H_R + E_w + E_b) \quad \text{for dizygotic twins of same sex}$$

$$= 0.49(\tfrac{1}{2}D_R + \tfrac{1}{4}H_R + E_w + E_b) \quad \text{for dizygotic twins of opposite sex}$$

$$= 0.51(\tfrac{1}{2}D_R + \tfrac{1}{4}H_R + E_w + E_b) \quad \text{for full-sibs.}$$

The last three have identical expectations on this model which assumes that they are subject to the same environmental sources of variation. Since the three correlations do not differ significantly we can accept that this is the case. It is, however, of interest to note that the non-significant differences between them fit a pattern in which the full-sibs appear to have been subjected to less environmental differences than dizygotic twins and dizygotic twins of different sexes subjected to less environmental differences than twins of the same sex. In the absence of significance, however, we may pool them. In effect we now have three equations for solving the four unknowns, the third being $\frac{1}{2}D_R + \frac{1}{4}H_R + E_w + E_b = 1.00$. We can therefore, estimate three quantities as proportions of the total variance

$$\frac{1}{2}D_R + 3E_b = 1.00$$
$$\frac{1}{4}H_R - 2E_b = 0.04$$
$$E_w = 0.04.$$

One conclusion that can be drawn from these estimates is that H_R, E_w and E_b must be small relative to D_R but to go beyond this we should need to test the adequacy of all possible two parameter models as described earlier. This is practicable with variances using weighted least squares techniques but is not with correlations.

Our aim in this section has been to illustrate the value and limitations of twin data and for this reason we have confined the discussion to twins and the only other kind of relationship (full-sib) that has the same expectation on a simple model. Twin data, however, are at their most powerful when supplementing the commoner types of relationships found in natural populations, (Eaves, 1975; Jinks and Fulker, 1970). But because they are more powerful, more complex sources of variation become amenable to analysis and sources of variation that do not normally arise in experimental populations reach significance and must be allowed for in any adequate model. These sources would, for example, include assortative mating, genotype-environment correlations, co-operation or competition between siblings and cultural transmission from parent to offspring. These developments are beyond the scope of our present treatment but they are described by Eaves *et al.* (1977).

33. Experimental analysis

The analysis of variation in a population becomes possible by experimental means in species where we can use controlled matings and raise the progenies in such a way that we can determine the impact on them of the effects of the non-heritable sources of variation. A number of experimental breeding programmes are then possible, of which the simplest is the use of biparental progenies produced by the mating of pairs of parents taken at random from the population, no parent being used more than once. With hermaphroditic plants, half the individuals would be used as males and half as females, and with species where the sexes are separate, equal numbers of males and females would be taken and mated in pairs taken at random. We should thus have a number of full-sib families which could all be made to comprise the same number, n, of individuals. Then from Table 53 we should have

$$\text{variance of family means } (V_{\bar{F}}) = \tfrac{1}{4}D_R + \tfrac{1}{16}H_R + E_b + \tfrac{1}{n}(\bar{V}_F)$$
$$\text{and mean variance of families } (\bar{V}_F) = \tfrac{1}{4}D_R + \tfrac{3}{16}H_R + E_w$$

where D_R and H_R are the parameters of the population from which the parents of the families were taken at random. $V_{\bar{F}}$ can obviously be corrected for $\frac{1}{n}(\bar{V}_F)$ to give an estimate of $\frac{1}{4}D_R + \frac{1}{16}H_R + E_b$. The analysis can, however, be taken further only if we make further assumptions or elaborate the design of the experiment. Thus if each family is divided into, say, halves and each half raised in separate randomized blocks we can obtain from the family × block interactions an estimate of $E_b + \frac{1}{n}(\bar{V}_F)$ and hence of E_b and $\frac{1}{4}D_R + \frac{1}{16}H_R$. Even so only if we could assume H_R to be zero would we be able to estimate D_R. The difficulty is that such an elaboration would still not provide enough statistics to estimate all the parameters. A further statistic is in fact necessary if the analysis is to be completed.

This further statistic might be sought in either of two ways. First we could use the parent/offspring covariance, but we should have to take steps to ensure that in doing so we were not introducing the further parameter E_b' and if it were necessary to raise parents and offspring in different environments, as for example, if they had to be grown in different years, their covariance might be biased by genotype × environment interaction. Given however, that a satisfactory estimate of the covariance could be obtained, it would supply us with a further statistic whose expectation is $\frac{1}{4}D_R$ and the analysis could be completed.

The second, and preferable, approach is to vary the design of the experiment so as to include not only families of full-sibs but new families standing in the half-sib relation to one another. This can be achieved by adopting the design often referred to as North Carolina (NC1), and involves the mating of each parent of one sex (usually for obvious reasons, the male) with a number of parents of the other sex, the group of individuals used of the second sex being a different one for each individual of the first sex. Thus Robinson *et al.* (1949) record an experiment with maize in which 48 plants used as males were each crossed on to 4 females, making a total of 4 × 48 = 192 females in all, both the males and the females to which each was crossed being taken at random from the population. This population was in fact the F_2 of a cross between two inbred lines, CI21 and NC7, but the experiment will serve to illustrate the use of the NC1 design which can be used just as well with any open-bred population as with an F_2: as we have noted earlier, an F_2 may properly be regarded as a randomly bred population but with the special condition that $u = v = \frac{1}{2}$ for all genes. Thus the only special feature of the results of Robinson *et al.* is that they will yield estimates of D and H rather than just D_R and H_R, since again as we have already seen D and H are the special cases of D_R and H_R where $u = v = \frac{1}{2}$ for all genes. We will, however, use D_R and H_R in our present analysis as a continuing reminder of the general applicability of the analysis.

The families produced by the 192 crosses were grown in 12 blocks, each block including the 16 families from the crosses of 4 males each to its 4 females. Each block was divided into 2 sub-blocks and all of the 16 families of the block were grown in each sub-block, randomization of the 16 families being carried out separately for the 2 sub-blocks. The data we will use relate to yield of grain, expressed as mean pounds per plot.

The analysis of variance is shown in Table 54. Each block includes 32 plots divided into two sub-blocks of 16 plots each. There is thus 1 df for

TABLE 54.

Analysis of variance of yield in maize (Robinson *et al.*, 1949)

Item	df	MS
Blocks	11	0.0153
Sub-blocks	12	0.0063
Male groups	36	0.0167 *
Families within groups	144	0.0069 *
Plots within families	178	0.0031 *
Sampling variance of plot means	250	0.0017

The analysis is in terms of plot means.

* Significant when tested against the appropriate error variance, which in all these cases is the MS immediately below.

the difference between sub-blocks, 15 for differences among the 16 families in the block and 15 for sub-block × family interaction. The first item is of little interest to us, but the second provides information about the effects of the genetical differences among the 16 families and the third item is a direct measure of the variance of the non-heritable component of variation in the family means. The 15 df for family differences are subdivisible into 3 for differences among the progenies of the 4 males and 3 × 4 = 12 for the differences among the progenies of the females mated to the same male, averaged over the 4 males of the block. This last item is clearly a measure of the variance of means of full-sib families, while the former measures the variance among the means of half-sib groups of families, since the 4 families tracing back to a single male each has a different mother and are therefore in the half-sib relationship to one another.

Since the 12 blocks are derived from 12 different sets each of 4 males and 16 females, we can pool corresponding items from all the blocks and find 1 × 12 = 12 df for sub-block differences, 3 × 12 = 36 for differences among the progenies of different males, 12 × 12 = 144 for differences among the females mated to the same male, and 15 × 12 = 180 df for the non-heritable component of variation of family means. Since, however, two plots failed in the experiment, their means were estimated by the standard missing plot technique and 2 df were lost from this total of 180 leaving 178 in the analysis. There are of course 11 df for differences among the 12 block totals, but, like the 12 df for sub-block differences, these are of little interest to us. Each plot contained 10 plants except in a few cases. The results were recorded as the mean yield per plant for each plot and an analysis of variance was carried out on a single plot basis. A further observation was, however, made. The mean variance of plants within plots was found from a sample of the plots used in this and another related experiment and used to derive an estimate of the sampling variance of the plot means, which is recorded as 0.0017 by Robinson *et al.* Where $\bar{V}_w$ is the mean variance within plots the sampling variance of the mean of plots of 10 plants would be $\frac{1}{10}\bar{V}_w$, but there were missing plants in a few plots and the divisor 10 was therefore replaced by 9.4 which is the harmonic mean of the actual numbers of plants in the plots.

The results of the analysis of variance require little comment. The non-heritable variation of plot means, estimated from the family × sub-block interaction, is clearly greater than the sampling variance of plot means arising from the variance of plants within plots. The MS for family ×

sub-block interaction must therefore be used for testing the MS between females within males which, if significant, must itself be used for testing the MS between males. Although the VR's are not large, with the high number of df available these two items are both significant when so tested and thus combine to provide evidence for genetical variation among the families. The differences between sub-blocks are not significant, while those between blocks are, but as already noted these items are of little interest for our present analysis and will be used no further.

The further analysis of the variation into the various heritable and non-heritable components can be carried out directly from the MS's in the analysis of variance set out in Table 54. This is in fact the approach used by Robinson *et al.* (and see M and J, pp. 226 et seq.). It is, however, somewhat easier to follow if we first find the variance of plot means, that of family means within male groups (i.e. within groups having a common male parent) and that between male group means, all of which are easily derivable from the MS's of Table 54. Since the analysis of variance was based on single plot observations, the variance of plot means within families is given directly by the MS for family × sub-block interaction. Each family included two plots, one in each sub-block, and the variance of family means within male groups is thus $\frac{1}{2}$ the MS between families within groups. Finally each male group includes four families each raised in two plots, and the variance of male group means thus becomes $1/(4 \times 2) = \frac{1}{8}$ of the MS between males. The variances so calculated are listed in Table 55, which also includes the mean variance within plots.

TABLE 55.

Components of variation of yield in the maize experiment

Variance of	Observed	Sampling correction	Corrected
Male group means (V_M)	0.002 09	$\frac{1}{4}V_F = 0.000\,86$	$0.001\,23 = \frac{1}{8}D_R$
Family means within groups (V_F)	0.003 45	$\frac{1}{2}V_P = 0.001\,57$	$0.001\,88 = \frac{1}{8}D_R + \frac{1}{16}H_R$
Plots within families (V_P)	0.003 13	$\frac{1}{9.4}V_{2S} = 0.001\,70$	$0.001\,43 = E_b$
Plants within plots (V_{2SR})	0.015 98	——	$0.015\,98 = \frac{1}{4}D_R + \frac{3}{16}H_R + E_w$

$D_R = 0.009\,84$ $H_R = 0.010\,40$ $E_b = 0.001\,43$ $E_w = 0.011\,57$

Since each plot mean has a sampling variance of 0.0017 as shown in Table 54, the variance of plants within plots is this sampling variance multiplied by 9.4, the harmonic mean number of plants per plot.

Now from Table 53 the mean variance within full-sib families is $\frac{1}{4}D_R + \frac{3}{16}H_R + E_w$ and the sampling variance this contributes to the variance of plot means is thus $\frac{1}{n}V_{2SR}$ where n is the harmonic mean of the number of individuals in the various families, here 9.4. The other component of the variance of plot means within families is E_b. We can thus find from the data in Table 55, $E_b = 0.003\,13 - 0.001\,70 = 0.001\,43$. Since each family mean is derived from two plots it will be subject to a sampling variance of $\frac{1}{2}$ the variance between plots within families, and if we were taking the analysis no further we should deduct $\frac{1}{2}$ the variance of plots within families (i.e. $\frac{1}{2}E_b + \frac{1}{2n}V_{2SR}$) to obtain the overall genetical component of variation between families. We have, however, sub-divided this variation into two parts, that between families within male groups (i.e. between progenies each with its own mother but having a common father) and that between male groups (i.e. between groups of families, each of which group comprises families with a common father). Before we can proceed further, therefore, we must ascertain how the genetical components divide up between these two sub-divisions of the variation.

We can obtain this partition of the genetical components by reference back to Table 51, which sets out the matrix of matings between the three genotypes of male and the three corresponding types of female in respect of the gene difference A-a. Now each row represents the families obtained by mating the various types of female with a constant, or single, male. In other words the row in the table are a model for our male groups. The expectation for the genetic part of the variance of means of male groups is thus given by the variance of row means and turns out to be $\frac{1}{2}u_a v_a [d_a + (v_a - u_a)h_a]^2$ which on summing over all relevant genes becomes $\frac{1}{8}D_R$. The expectation for the genetic part of the variance of family means within male groups is similarly given by the mean variance of families within rows and this is found to be $\frac{1}{2}u_a v_a [d_a + (v_a - u_a)h_a]^2 + u_a^2 v_a^2 h_a^2$ which on summing over all relevant genes becomes $\frac{1}{8}D_R + \frac{1}{16}H_R$. When summed these two variances give

$$\tfrac{1}{8}D_R + \tfrac{1}{8}D_R + \tfrac{1}{16}H_R = \tfrac{1}{4}D_R + \tfrac{1}{16}H_R$$

which is, of course, the expectation we have already found for the overall variance of means of biparental families. The NC1 mating system has thus enabled us to break the overall variance of family means into two recognizable parts having different expectations in terms of our parameters and so add a further equation for the estimation of the parameters.

Returning to our analysis, we note that the means of families within male groups are each based on two plots. Their variance will thus have an expectation of $\frac{1}{8}D_R + \frac{1}{16}H_R + \frac{1}{2}E_b + \frac{1}{2n}V_{2SR}$ allowing us to estimate $\frac{1}{8}D_R + \frac{1}{16}H_R$ as $0.003\,45 - \frac{1}{2}(0.003\,13) = 0.001\,88$. Since the male groups each include four families their means will be subject to a sampling variance of one-quarter the variance of individual family means. Their expectation for the variance of male group means is thus $\frac{1}{8}D_R + \frac{1}{4}(\frac{1}{8}D_R + \frac{1}{16}H_R + \frac{1}{2}E_b + \frac{1}{2n}V_{2SR})$ and we can estimate $\frac{1}{8}D_R$ by deducting one-quarter the variance of family means within male groups from the variance of male group means, giving $\frac{1}{8}D_R = 0.002\,09 - \frac{1}{4}(0.003\,45) = 0.001\,23$. We now have the estimates $\frac{1}{8}D_R + \frac{1}{16}H_R = 0.001\,88$ and $\frac{1}{8}D_R = 0.001\,23$ giving $D_R = 8 \times 0.001\,23 = 0.009\,84$ and $H_R = 16 \times (0.001\,88 - 0.001\,23) = 0.0104$.

Finally we note that the variance of individuals within families is $\frac{1}{4}D_R + \frac{3}{16}H_R + E_w = 0.015\,98$ and now having estimates of D_R and H_R we can complete the analysis by finding

$$E_w = 0.015\,98 - \tfrac{1}{4}(0.009\,84) - \tfrac{3}{16}(0.010\,40) = 0.001\,16.$$

The estimates of the four parameters D_R, H_R, E_w and E_b are assembled at the foot of Table 55. Since there were only four equations (provided by the variances of male group means, of family means within male groups, of plot means within families and of individuals within plots respectively) the solutions give perfect fit estimates of the parameters and we therefore have no test of adequacy of the model: at least one more equation, whose provision would require the experiment to be further elaborated in an appropriate way, would be needed for such a test of adequacy.

Various more elaborate experimental designs have been proposed from time to time, and have indeed been used in practice in a limited number of cases. There is, for example, the design often referred to as North Carolina 2, in which a number of male and female parents are used, but with every male mated to every female. This yields a quasi-diallel set of crosses, resembling the diallel in that every male genotype is mated to every female, and of course vice versa; but differing from it in that (a) the male parents and female parents are separate samples from the population of genotypes, there being no necessary correspondence between them in either genotype or number, and (b) being samples from an open bred population, the parents are not fully homozygous as are the parents of the diallels we discussed in Chapter 4. The data from an NC2 experiment can nevertheless be analysed like a diallel, although for reason (b) above, they will not yield the same estimates of the genetical parameters as a true diallel. Thus, the variances of the means of both the male and the female arrays yield estimates of $\frac{1}{8}D_R$, and not of $\frac{1}{4}D_R$ as with a true diallel, and similarly the term for interaction of male and female parents in the simple analysis of variance of the quasi-diallel table depends on $\frac{1}{16}H_R$ not $\frac{1}{4}H_R$ as in the true diallel. Finally, the mean variance within families has a genetical component, $\frac{1}{4}D_R + \frac{3}{16}H_R$ in a quasi-diallel whereas in a true diallel this variance within families is wholly non-heritable. Since this design yeilds two estimates of D_R, from the means of male and female arrays respectively, it affords in principle a test of adequacy of the model, but it will clearly be more a test of the assumption that male and female parents contribute equally to the phenotype of the progeny, i.e. that there are, for example, no maternal effects, than of anything else.

Where a number of inbred, homozygous lines are available from the population, or are otherwise readily made from it, a true diallel experiment may be carried out and analysed in the normal way. Appropriate sets of homozygous lines will however seldom be available, although such a set has been used in at least one case. Where analysis can be carried out by such a true diallel experiment, it will afford a better test of adequacy of the model and will yield more informative estimates of the parameters in the sense that their standard errors will be lower from an experiment involving a given number of individuals, than will any of the other designs, just as NC2 is more informative than NC1 (M and J, pp. 241–3). A true diallel, however, demands a suitable sample of homozygous lines, and even an NC2 requires the capacity for producing a series of different progenies from a single female by controlled matings with successive males. Such a controlled multiplicity of matings is more likely to be possible with plants than with animals, where indeed the possibilities must commonly be restricted to the NC1 design. In general the choice of design will be governed more by the biological possibilities of the species than anything else. Also because the analysis of NC1 experiments depends on the partitioning of variances, and variances whose genetical components involve D_R and H_R with such low coefficients as 1/8 and 1/16, such experiments must be large, involving large numbers of individuals and hence demanding of resources to carry out, if they are to yield informative estimates of the genetical components.

34. Complicating factors

The assumptions on which is based the model we have used in the genetical analysis of populations are (a) that the genes, both allelic and non-allelic, are distributed independently of one another in the population under analysis and (within the limits imposed by the mating system used) in the progenies on which are based the observations used in the analysis, and (b) that the genes display neither non-allelic interaction nor genotype × environment interaction in expressing their effects. The assumption of independence of gene distribution is primarily the assumption of random mating: linkage will have little effect in a randomly mating population unless the forces of selection impinging on the population are such as to produce a marked linkage disequilibrium. The assumption of random mating does not always hold good. We have already seen that there is assortative mating (that is a phenotypic correlation between mates) in man and it is known that mating can depart from randomness in populations of other animal species also. Indeed anything that affects the time of sexual maturity or mating behaviour and choice can prospectively lead to non-random mating. In plants a variety of mechanisms are known to affect mating, some leading to an excess of self-mating and others virtually to exclusive cross-mating. The latter may be regarded as a means of ensuring effectively random mating in respect of all the genes except those governing the mechanism itself (see Mather, 1973). The former by encouraging self-mating must generally lead to marked departures from randomness in the direction of inbreeding and hence to proportions of homozygotes in excess of those expected from the Hardy-Weinberg equilibrium in respect of any genes that vary in the population.

Assortative mating is the preferential coming together of individuals in mating pairs on the basis of similarity (or, in negatively assortative mating, of dissimilarity) of their phenotypes. Inbreeding is the preferential coming together of individuals in mating pairs on the basis of closer than average family, and hence genetic, relationship. Inbreeding may be held to imply a form of assortative mating; but the distinction between them is nevertheless an important one, as their consequences are not the same. They differ in several ways. Inbreeding will tend to raise the proportion of homozygotes in the population and if sufficiently close will lead to complete homozygosis apart from the effect of recurrent mutation.

Furthermore it will do so for all the genes in the nucleus, with the result that, as in Johannsen's beans, the population will consist of a mixture of true-breeding lines. Assortative mating on the other hand, depending as it does only on phenotypic similarity, will be affected by non-heritable agencies as well as by heritable: it will affect the distribution of the genes mediating the character in question, but it need not lead to any marked increase in homozygosis, even where the contribution of non-heritable agencies is small. Indeed it will not result in any significant rise in the proportion of homozygotes where the variation in the expression of the character in question is mediated by a reasonably large number of gene-differences whose effects are not grossly dissimilar in magnitude. Thus the consequences of assortative mating and inbreeding will appear in different ways in respect of continuous variation. Because of the association of non-allelic genes of similar effect to which it leads, assortative mating raises the contribution of D_R to the variation of the character in the population, while in so far as it does not lower the proportion of heterozygotes, it leaves the contribution of H_R unchanged. Because it raises the proportion of homozygotes, inbreeding also raises the contribution of D_R to the variation, but because of the concommitant reduction in the proportion of heterozygotes, the contribution of H_R is correspondingly lowered. With complete inbreeding H_R vanishes entirely from the composition of the variation.

Where assortative mating is operative, it can be accommodated by the approach due to Fisher (1918) to which we have already made a brief reference, and which has been illustrated further in its analytical situation by Jinks and Fulker (1970). Where inbreeding is complete it is easily accommodated in the analysis. The population then consists of nothing but homozygotes in the proportions u AA:v aa, and its variance will be $D_P + E_w + E_b$, where $D_P = \mathrm{S}[4u_a v_a d_a^2]$ as shown when we were considering the variance of the homozygous parents of a diallel in Section 18. Where inbreeding is only partial the situation is more complex involving D_R, H_R and f, the inbreeding coefficient, as well as D_P. The analysis then becomes correspondingly complicated.

Turning to interactions, the presence of genotype × environment interaction is easy to detect by a comparison of the variance of the population over two or more environments. If the simple model assuming no such interaction is adequate, the variances of the population will be homogeneous: any significant heterogeneity of their variances will show that genotype × environment interaction must be taken into account. Kearsey (1965) has reported an analysis of the variation in flowering time of a randomly bred population of the poppy, *Papaver dubium*, which he carried out using a number of experimental designs, two of which were NC1 and NC2. He sowed samples of each of the experimental progenies that he used in the analysis of the population, at two different times, so making it possible to compare the variances they yield when grown in the two different environments experienced by plants raised at two different periods of the year. The mean variances of the families following the two sowings are shown for both his NC1 and NC2 experiments in Table 56. Each of these four MS are based on over 320 df, and it is clear

TABLE 56.

Variation in flowering time of a population of poppies (Kearsey, 1965)

	Sowing	Experiment NC1	Experiment NC2	Mean	Ratio 1/2
$\bar{V}_F$	1	36	49	42.5	2.02
	2	19	23	21.0	
D_R	1	45	30	37.5	3.13
	2	10	14	12.0	
H_R	1	76	159	117.5	2.67
	2	46	42	44.0	
E_W	1	11	10	10.5	1.05
	2	8	12	10.0	

that in both experiments the mean variance of families $\bar{V}_F$, which is of course $\frac{1}{4}D_R + \frac{3}{16}H_R + E_w$, is lower with sowing 2 than with sowing 1. His data allow estimates to be obtained of D_R, H_R and E_w from both the NC1 and NC2 experiments, and these are also set out for both sowing times in the table. If we take their averages over the two experiments both of the genetical parameters are about three times as high after sowing 1 than after sowing 2, but E_w hardly changes between sowings. The difference in the variation between environments is thus unlikely to be one that can be scaled out by transforming the metric on which the character has been measured, and we must conclude that expressions of the genes mediating the variation in flowering time are changing markedly with the change in environment.

Interaction between genotype and environment is relatively simple to detect. That between non-allelic genes, on the other hand, is difficult. As in the descendants of a cross between true-breeding lines (Section 21), the effects of non-allelic interaction on the genetical component of variation in a randomly breeding population are two-fold (Mather, 1974).

First, the terms in D_R and H_R have added to them terms in I_R, J_R and L_R. These terms appear with the same coefficients as do I, J and L in the corresponding variances and covariances of F_2 and its descendants. They are set out in the upper part of Table 57. Secondly, the non-allelic inter-

TABLE 57.

Non-allelic interaction in randomly breeding populations (Mather, 1974)

$$V_R = \tfrac{1}{2}D_R + \tfrac{1}{4}H_R + \tfrac{1}{4}I_R + \tfrac{1}{8}J_R + \tfrac{1}{16}L_R + E_w + E_b$$
$$V_{1SR} = \tfrac{1}{4}D_R + \tfrac{1}{16}H_R + \tfrac{1}{16}I_R + \tfrac{1}{64}J_R + \tfrac{1}{256}L_R + E_b + \tfrac{1}{n}V_{2SR}$$
$$V_{2SR} = \tfrac{1}{4}D_R + \tfrac{3}{16}H_R + \tfrac{3}{16}I_R + \tfrac{7}{64}J_R + \tfrac{15}{256}L_R + E_w$$
$$W_{SR} = \tfrac{1}{4}D_R + \tfrac{1}{16}H_R + \tfrac{1}{16}I_R + \tfrac{1}{64}J_R + \tfrac{1}{256}L_R$$
$$W_{HSR} = \tfrac{1}{8}D_R + \tfrac{1}{64}I_R$$
$$W_{POR} = \tfrac{1}{4}D_R + \tfrac{1}{16}I_R$$

where

$$D_R = S_a[4\,\Pi_a\{[d_a + 2S_b(\Pi_b j_{ab}) + S_b(\Delta_b i_{ab})] - \Delta_a[h_a + S_b(\Delta_b j_{ba}) + 2S_b(\Pi_b l_{ab})]\}^2]$$
$$H_R = S_a[16\,\Pi_a{}^2\{h_a + S_b(\Delta_b j_{ba}) + 2S_b(\Pi_b l_{ab})\}^2]$$
$$I_R = S_{ab}[16\,\Pi_a\Pi_b\{i_{ab} - \Delta_b j_{ab} - \Delta_a j_{ba} + \Delta_a\Delta_b l_{ab}\}^2]$$
$$J_R = S_{ab}[64\,\Pi_a\Pi_b\{\Pi_b(j_{ab} - \Delta_a l_{ab})^2 + \Pi_a(j_{ba} - \Delta_b l_{ab})^2\}]$$
$$L_R = S_{ab}[256\,\Pi_a{}^2\,\Pi_b{}^2\,l_{ab}^2]$$

and

S_a = summation over all genes
S_b = summation over all genes interacting with A-a
S_{ab} = summation over all pairs of interacting genes
$\Pi_a = u_a v_a$ and $\Delta_a = u_a - v_a$

action changes the definitions of D_R and H_R in a randomly breeding population, just as it changes those of D and H in F_2 although in a more complex way. Indeed D_R and H_R are now affected by the i, both j's and l for each pair of interacting genes, and not just by j and l respectively as are D and H. The definitions of D_R, H_R, I_R, J_R and L_R are also set out in Table 57. They are very complex, but reduce to the simpler expressions for the D, H, I, J and L of F_2 when all $u = v = \tfrac{1}{2}$.

It will be seen from the table that I_R appears whenever D_R is present in a variance or covariance, while J_R and L_R appear whenever H_R is present. It is thus difficult to separate I_R from D_R and J_R and L_R from H_R and this necessarily aggravates the problem we have already met in separating the E components of variation from D_R and H_R. In an F_2 and its descendants we can detect interaction by the changes in value of D and H with generations, but this approach is not available to us with a randomly breeding population since all the variances and covariances we obtain from the population itself or from the test matings made in it are the equivalent of first generation statistics, and if we go on to produce the equivalent of second or later generations we run into the difficulties we have already seen to arise in partially inbred populations. Nor, for reasons which we saw at the beginning of this chapter, can we use that most powerful means of all for detecting non-allelic interaction, the scaling test. Thus our estimates of D_R and H_R are subject to distortion both by the difficulty of separating I_R from D_R and J_R and L_R from H_R, and by the direct impact of the i's, j's and l's on the D_R and H_R themselves, while at the same time the presence of the interaction causing the distortion may pass undetected, save in special cases. This is a subject worthy of more attention than it has yet received.

35. Heritability

The proportion that the heritable variation constitutes of the total phenotypic variation of a character in a population is commonly referred to as the heritability of that character. The heritability is generally denoted by h^2, but to avoid confusion with h and h^2 as we have been using them, we will here denote it by T. A distinction is further drawn between what are termed the 'narrow' heritability and the 'broad' heritability. The former is the proportion that the additive genetic variation constitutes of the total variation, and the latter is the proportion that all the heritable or genotypic variation constitutes of the total. Thus where both additive and dominance variation are present (but leaving aside non-allelic interaction) the narrow heritability in a population is $T_n = \frac{1}{2}D_R/(\frac{1}{2}D_R + \frac{1}{4}H_R + E_w + E_b)$ while the broad heritability is $T_b = (\frac{1}{2}D_R + \frac{1}{4}H_R)/(\frac{1}{2}D_R + \frac{1}{4}H_R + E_w + E_b)$. Where dominance variation is absent $T_n = T_b = \frac{1}{2}D_R/(\frac{1}{2}D_R + E_w + E_b)$. It should be noted that non-allelic interaction like dominance can change T_b without altering T_n to a corresponding extent.

The heritability, and particularly the narrow heritability, T_n, provides a convenient summary of the situation with regard to the distribution of variation between the genetic and the non-genetic within the population. It is easily measured as the ratio that twice the parent/offspring covariance ($W_{POR} = \frac{1}{4}D_R$) bears to the variance of the individuals in the population, provided that E_b' can be shown to be negligible or can be made negligible or can be measured and deducted from W_{POR} to leave a direct estimate of $\frac{1}{4}D_R$. Furthermore, once we know the value of T_n it can be used to predict the response of the population to certain types of selection. Thus if we select that group of individuals which has a greater expression of the character than the remaining group of unselected individuals and then breed them together, the mean expression of the offspring so obtained will exceed that of the population by $R = T_n S$ where R is referred to as the response to selection and S, the intensity of selection, is the amount by which the mean of the selected parents exceeds that of the population (see Falconer, 1960). As Falconer points out, this prediction of selective response will hold good in detail only where a number of other conditions apply, for example, that there is no non-allelic interaction and the scale of measurement is adequate. In any case the predictions can be expected to be valid only in the short-term, since response to selection must itself imply changes of gene frequency, including some gene fixation. Nevertheless predictions of this kind have proved to hold good, at least to a first approximation, in a high proportion of cases.

The uses to which the concept of heritability can be put should not, however, blind us to its limitation. These stem ultimately from two of its features. In the first place it is a ratio, in the case of T_n the ratio of the additive genetical variation to the total phenotypic variation of the population. It depends therefore not just on the amount of heritable variation in the population, but also on the amount of non-heritable. The heritability can be raised not only by injecting more genic variation into the population but also by making more stable the environment in which the individuals are raised and develop. Equally it can be lowered by raising the non-heritable variation as well as by reducing the heritable. Thus, while the heritability may be a convenient summary of the situation for some comparisons or uses, it can never give as clear and informative a picture as the estimates of the components of variation, D_R, H_R and E. Given such estimates we can easily construct T_n or T_b whichever we need, should we need it, and at the same time we have comprehensive information which provides an understanding beyond anything to be obtained from heritabilities and their comparison.

The second limitation of the concept of heritability stems from the properties of the genetical components of variation, especially D_R, of which it is compounded. As we have already noted, since $D_R = \mathrm{S}\, 4uv[d + (v-u)h]^2$ it cannot give us information about the genetical potentialities of the population in the way that $D = \mathrm{S}(d^2)$ can do for the descendants of a cross between two inbred lines. The value of D_R not only varies with the gene frequencies as a result of the general factor uv that it contains for each gene difference, but it also depends on the term $(v - u)h$ which is included with d. Now if the more common of two alleles is dominant, $v < u$ when h is positive and $v > u$ when h is negative. In either case $(v - u)h$ will be negative and $d + (v - u)h$ will be less than d. In the same way when the less common allele is dominant, $(v - u)h$ will be positive and $d + (v - u)h$ will be greater than d. We can illustrate the effect of this relationship by reference to the data of Robinson *et al.* (1949) on yield in maize, which we analysed in Section 33. Although we used the data there to illustrate the analysis of a population by means of the NC1 experimental design, the results were in fact derived from an F_2 where of course all $u = v = \frac{1}{2}$, giving $D_R = D = S(d^2)$ and $H_R = H = S(h^2)$. We found $D_R = D = 0.009$, $H_R = H = 0.010$ and $E_w + E_b = 0.013$. Approximating these findings by setting $D = H = E_w + E_b = 0.01$ for ease of presentation, we note that if all the genes in the system are alike in their effects $h/d = \sqrt{(H/D)} = 1$ and $d = h$. Then assuming that u and v are the same for all genes we can calculate T_n and T_b for any gene frequency that we choose. The relations of T_n and T_b to u, so obtained, are shown in Fig. 21, from which we see that T_n becomes increasingly small as u

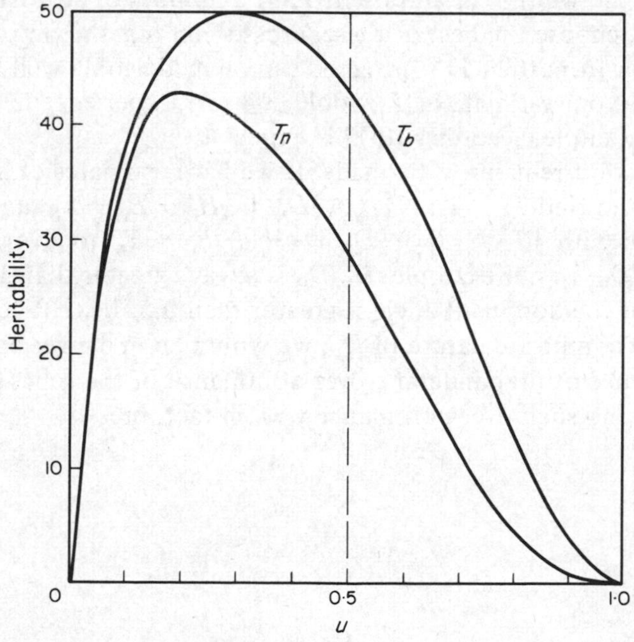

Fig. 21. Effect of gene frequency, u, on the narrow (T_n) and broad (T_b) heritabilities, in %, in a randomly breeding population, where $Sd^2 = Sh^2 = E = 1$. d, h and u are assumed to be the same for all gene pairs.

increases above 0.5, and in particular becomes relatively very small as u rises to 0.8 or more. When, however, $u < 0.5$, T_n can rise to 3/2 the value it has at $u = v = \frac{1}{2}$, before falling away towards 0 as u approaches 0. Thus when $u > 0.5$, T_n will always underestimate the fixable genetic variation and will grossly underestimate it as u approaches 1. When $u < 0.5$, T_n can materially overestimate the fixable genetic variation until u gets fairly close to 0. If we had taken $h = -d$, which is also consonant with the data, the same pair of curves would have been obtained but with $v = 1 - u$ replacing u along the abscissa.

In both cases the abscissa is the frequency of the dominant-allele and T_n always gives an underestimate of the fixable genetic variation when the dominant gene is the more common; although it generally overestimates it when this allele is the less common. Such evidence as we have suggests that the dominant allele tends to be the more common in populations. We must expect therefore that although T_n may tell us how the population will respond to simple mass selection, it will underestimate the changes that can be obtained if we set about our breeding programme in a different way. If, for example, instead of applying mass selection to the population, we first of all raise from it a number of at least partially inbred lines, choose the best of these, cross them together in pairs and select further from their F_2's, progress can be made going well beyond anything that our estimate of T_n would suggest. Experience in breeding maize, for example, accords with this expectation.

One last point remains to be made. If we have estimates of both T_n and T_b, we can find $T_b - T_n = \frac{1}{4}H_R/(\frac{1}{2}D_R + \frac{1}{4}H_R + E_w + E_b)$ and this can be compared with $\frac{1}{2}T_n = \frac{1}{4}D_R/(\frac{1}{2}D_R + \frac{1}{4}H_R + E_w + E_b)$ to give us an estimate of H_R/D_R. In our example, H_R/D_R is always greater than 1 when the frequency of the dominant allele is greater than 0.5. If we failed to remember the composite nature of D_R, we would be in danger of taking this as evidence of preponderant over-dominance of the genes in the population, when no such over-dominance was, in fact, present.

8

Genes and effective factors

36. Estimating the number of segregating genes

In the absence of non-allelic interaction the mean phenotypes of two true-breeding lines may, as we have seen in Chapter 3, be represented as $m + [d]$ and $m - [d]$ respectively, where m is the mid-parent value and $[d]$ is the sum of the d increments of all the genes in which the lines differ. Sign is taken into account in finding $[d]$ to accommodate the association in the two lines of the $-$ alleles at some loci with the $+$ alleles at others. Where, however, the $+$ alleles at all the k loci in which the lines differ, are associated in one parent and all the $-$ alleles in the other, $[d] = d_a + d_b \cdots d_k = \mathrm{S}(d)$ and this becomes kd where all the gene differences are of equal effect, that is $d_a = d_b = \cdots = d_k = d$. Thus with complete association of like alleles and with all the gene differences having equal effects the mean phenotypes of the two lines will differ by $2\mathrm{S}(d) = 2kd$. Now in the absence of linkage $D = \mathrm{S}(d^2) = kd^2$ and if we divide the square of half the parental difference by D we find

$$(kd)^2/D = k^2 d^2/kd^2 = k$$

and we have an estimate of k, the number of genes in which the two lines differ.

In arriving at this estimate of k we have made four assumptions, that:

(a) there is no non-allelic interaction,

(b) the gene differences are of equal effect,

(c) there is complete association of like alleles in the parents,

(d) there is no linkage of the genes.

What are the consequences on the estimate of k if these assumptions fail?

Taking non-allelic interaction first, it will be recalled from Section 20 that when allowance is made for such interaction the means of the two parental lines become $m + [d] + [i]$ and $m - [d] + [i]$. So, half the parental difference is still $[d]$, and no complication is introduced into the numerator of the fraction which yields our estimate of k. Turning to the denominator, however, we note that $D = S(d_a + \frac{1}{2}Sj_a)^2$ in F_2 and S_3 and it will exceed $S(d_a^2)$ or fall short of it according to the preponderant sign of the j's, and by an amount which will depend also on the extent and magnitude of this interaction (see Section 21). The estimate of k can thus be biased upwards or downwards by j interaction. If we have the data for estimating D in more than one generation we may be able to correct it for the effect of the interaction, since in F_3, $D = S(d_a + \frac{1}{4}Sj_a)^2$ and in F_4 it changes further to $S(d_a + \frac{1}{8}Sj_a)^2$ so allowing us to extrapolate to $S(d_a^2)$. An extensive set of observations would be necessary for such a procedure and no attempt has yet been made to find k in the known presence of non-allelic interaction.

Turning next to the assumption of equality of gene effects, we note that if these effects are not in fact equal we can define $\bar{d}$ as their average and then write $d_a = \bar{d}(1 + \alpha_a)$, $d_b = \bar{d}(1 + \alpha_b)$ and so on. It can then be shown that our estimate of k becomes $\hat{k} = k/(1 + V_\alpha)$ where V_α is the variance of α or equally the variance of $d/\bar{d}$ (see M and J, p. 309). Thus inequality of the gene effects must always lead to an underestimate of k.

To take an example, where there are three gene differences of equal effect, d being 2 for each of them ($d_a = d_b = d_c = 2$) with the + alleles all in one parent and the − alleles in the other (shown as $\frac{2,\ 2\ \ 2}{-2, -2, -2}$ in Table 58). $[d] = 2 + 2 + 2 = 6$ and $D = S(d^2) = 12$, giving $\hat{k} = 6^2/12 = 3$, which of course equals the true k. If however, we have three genes of unequal effects, with $d_a = 3$, $d_b = 2$, $d_c = 1$, again with complete association of like alleles in the parents (shown as $\frac{3,\ 2,\ 1}{-3, -2, -1}$ in Table 58), $[d] = 3 + 2 + 1 = 6$ as before, but $D = 3^2 + 2^2 + 1^2 = 14$ giving $\hat{k} = 6^2/14 = 2.57$, so underestimating k. In this case $\bar{d} = \frac{1}{3}(3 + 2 + 1) = 2$ and $d_a = 2(1 + \frac{1}{2})$, $d_b = 2(1 + 0)$ and $d_c = 2(1 - \frac{1}{2})$ giving $\alpha_a = \frac{1}{2}$, $\alpha_b = 0$, $\alpha_c = -\frac{1}{2}$ and $V_\alpha = \frac{1}{3}[\frac{1}{2}^2 + 0^2 + (-\frac{1}{2})^2] = \frac{1}{6}$. Then $\hat{k} = k/(1 + V_\alpha) = 3/(1 + \frac{1}{6}) = 2.57$ as already found.

Incomplete association of like alleles also leads to an underestimate, and generally a much greater underestimate, of k, since $[d]$ is necessarily less than $S(d)$. If we write $S(d_+)$ for the summed effects of the genes, whose + alleles are present in the larger parent and $S(d_-)$ for the summed effects of those whose − alleles are also present in that parent, $[d] = S(d_+) - S(d_-) = S(d) - 2S(d_-)$ and we can obtain a measure of the

TABLE 58.

The consequences of inequality of gene effects and incomplete association of like alleles for the estimate of the number of gene differences.

[Note: The effects of the three gene differences and the distribution of alleles between the parents are shown in the left-hand column. Thus, for example, $\frac{2\ \ 2\ \ 2}{-2\ -2\ -2}$ indicates that all gene differences are of equal effect (all $d = 2$) with the + alleles concentrated in one parent and the − alleles in the other; while $\frac{3\ -2\ \ 1}{-3\ \ 2\ -1}$ indicates gene differences of unequal effect ($d = 3$, 2 and 1 for them respectively) with the − allele of the second gene associated with the + alleles of the other two.]

	Assumptions		$[d]$	r	D	V_α	$\hat{k}$
	Equal effects	Complete association					
$\frac{2\ \ 2\ \ 2}{-2\ -2\ -2}$	v	v	6	1	12	0	3.00
$\frac{3\ \ 2\ \ 1}{-3\ -2\ -1}$	f	v	6	1	14	$\frac{1}{6}$	2.57
$\frac{2\ \ 2\ -2}{-2\ -2\ \ 2}$	v	f	2	$\frac{1}{3}$	12	0	0.33
$\frac{3\ \ 2\ -1}{-3\ -2\ \ 1}$	f	f	4	$\frac{2}{3}$	14	$\frac{1}{6}$	1.14
$\frac{3\ -2\ \ 1}{-3\ \ 2\ -1}$	f	f	2	$\frac{1}{3}$	14	$\frac{1}{6}$	0.29
$\frac{3\ -2\ -1}{-3\ \ 2\ \ 1}$	f	f	0	0	14	$\frac{1}{6}$	0.00

$\hat{k} = [d]^2/D = kr^2/\bar{d}^2(1 + V_\alpha)$. In all cases $k = 3$ and $\bar{d} = 2$; v = assumption valid; f = assumption invalid

degree of association by setting $r_d = [S(d) - 2S(d_-)]/S(d)$. This will of course be 1 when association is complete and 0 when dispersion of like alleles between the two parents is at its effective maximum. The estimate of k thus becomes $\hat{k} = [d]^2/D = [r_d S(d]^2 D = kr_d^2$, which must tend to be an underestimate since r_d lies between 1 and 0. If the assumptions of equal gene effects and complete association fail simultaneously, it can be shown (M and J, p. 310) that $\hat{k} = kr_d^2/(1 + V_\alpha)$, with the inequality of effects and incompleteness of association reinforcing one another in reducing k.

We can illustrate the consequences of incomplete association by reference to the basic example already used to illustrate the consequences of inequality of effects. With three genes of equal effect, all $d = 2$, but with two of their + alleles associated with the − allele of the third ($\frac{2,\ 2,\ -2}{-2,\ -2,\ 2}$ in Table 58), $[d] = 2 + 2 - 2 = 2$ while $D = 12$ as before. Then $\hat{k} = 2^2/12 = 0.33$, whereas of course k still is 3. Looking at it in the alternative way $r_d = \frac{1}{6}[6 - (2 \times 2)] = \frac{1}{3}$ and $\hat{k} = kr_d^2 = 3 \times (\frac{1}{3})^2 = 0.33$. With the genes of unequal effects, $d_a = 3, d_b = 2, d_c = 1$ there are three possible distributions between the parents as shown in Table 58, and each gives its own characteristic underestimate of k. Thus $\frac{3,\ 2,\ -1}{-3,\ -2,\ 1}$ has $[d] = 3 + 2 - 1 = 4$ and $D = 14$ as in the earlier example, giving $\hat{k} = 4^2/14 = 1.14$. Put the other way, $r_d = \frac{1}{6}[6 - (2 \times 1)] = \frac{2}{3}$ with $V_\alpha = \frac{1}{6}$ as found in the earlier example, so giving $\hat{k} = kr_d^2/(1 + V_\alpha) = [3 \times (\frac{2}{3})^2]/(1 + \frac{1}{6}) = 1.14$. The values of $[d]$, D, r_d, V_α and $\hat{k}$ are also shown in the table for the other two possible distributions of the alleles.

37. Consequences of linkage: effective factors

The fourth assumption we made in arriving at our estimate of the number of genes was that the genes were unlinked, and we must now consider the effects of linkage. Now, as we saw in Section 22, linkage has no effect on family means, provided there is no non-allelic interaction of the linked genes, but it does however affect D, which no longer is $S(d^2)$ but includes terms in $d_a d_b$ and p, the recombination value. We can illustrate the consequences of this change in D for our estimate of k by considering the case of two genes A-a and B-b, where $d_a = 3$, $d_b = 1$ and the recombination value is p.

Two distributions are possible of the genes between the parent lines. In one, the like alleles are associated in the two parents which are thus AABB and aabb or $\frac{3,\ 1}{-3,\ -1}$ in the notation of the previous section. In the other the genes are dispersed, the parents being AAbb and aaBB or $\frac{3,\ -1}{-3,\ 1}$ in the same notation. The associated distribution will lead to oupling linkage in F_1 and so may be denoted by C, while the dispersed istribution will give repulsion linkage and so may be denoted by R. Vith the C arrangement $[d] = d_a + d_b = 3 + 1 = 4$, and $D = d_a^2 + d_b^2 + d_a d_b(1-2p) = 3^2 + 1^2 + 2.3.1(1-2p) = 16 - 12p$, and with the R rrangement $[d] = 3 - 1 = 2$ and $D = d_a^2 + d_b^2 - 2d_a d_b(1-2p) =$ $+ 12p$. We thus find, $\hat{k}_C = 4^2/(16-12p)$ and $\hat{k}_R = 2^2/(4+12p)$. In he absence of linkage $p = 0.5$ and $\hat{k}_C = 4^2/10 = 1.6$, and $\hat{k}_R = 2^2/10 =$).4 the departure from the true value of $k = 2$ being due partly, as with $\hat{k}_C$, to the inequality of d_a and d_b, but now chiefly to the dispersion of ike alleles between the parents.

When however, linkage is complete $p = 0$, and $\hat{k}_C = 4^2/16 = 1$ while $\hat{k}_R = 2^2/4 = 1$. No matter whether the genes are in the coupling or repulsion arrangement we now arrive at the conclusion that there is but one gene difference between the parents, as indeed we should since two completely linked genes are effectively a single unit of inheritance. The lifference between the two cases is in the effects produced by the two alleles of the composite unit; with coupling they are AB and ab giving the components $[d] = 3 + 1 = 4$ and $D = (3+1)^2 = 16$, while with repulsion they are Ab and aB giving $[d] = 3 - 1 = 2$ and $D = (3-1)^2 = 4$.

With two genes linked but some recombination between them, values of k are obtained intermediate between 1.0 and 1.6 for coupling and between 1.0 and 0.4 for repulsion, as illustrated in Fig. 22 where $\hat{k}$ is plotted

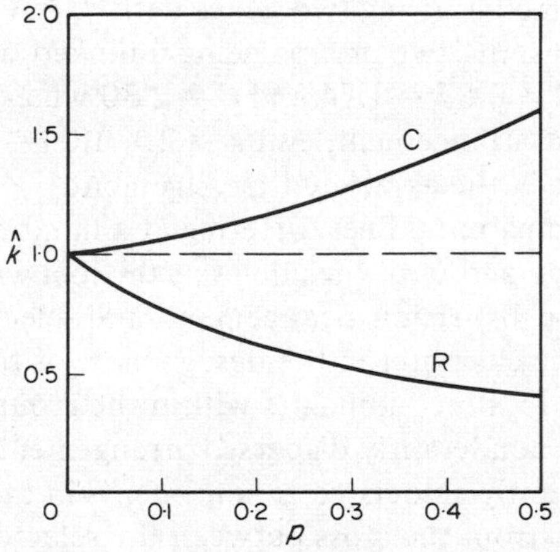

Fig. 22. Effect of linkage on the estimate, $\hat{k}$, of the number of units of inheritance where two genes, with $d_a = 3$ and $d_b = 1$ show the recombination frequency p. C indicates the coupling (that is, like alleles associated) and R the repulsion (that is, like alleles dispersed) arrangements of the genes.

against p. This is of course to be expected but we should note that with tight linkage $\hat{k}$ lies close to 1, and even with p as high as 0.1, $\hat{k}$ is still close to 1, especially with coupling where it is 1.08, although even with repulsion it has fallen only to 0.77. Thus even where recombination occurs, the two genes still appear more like a single unit of inheritance than like two, unless the linkage is loose and recombination fairly frequent. Where linkage is reasonably tight therefore we are estimating not the number of genes but the number of effective units of inheritance or effective factors as they are termed. We should note further that with reasonably tight linkage $\hat{k}$ is much the same whether measured from the coupling or the repulsion cross. Thus with $p = 0.05$, $\hat{k}_C = 1.04$ and $\hat{k}_R = 0.87$. The difference between the two cases lies not so much in the number of effective factors as in the average effect of that factor: with coupling $[d]_C = 4$ and $\bar{d}_C = [d]_C/\hat{k}_C = 4/1.04 = 3.85$ while with repulsion $[d]_R = 2$ and $\bar{d}_R = [d]_R/\hat{k}_R = 2/0.87 = 2.30$. This is of course a very simple example that we have taken for illustrative purposes. Clearly however, the same principle will hold where a greater number of genes are linked and so aggregated into a single effective factor. At the same time the number of possible arrangements of the genes in relation to one another is much greater and the change in the effect of the factor from the most dispersed to the most associated will be correspondingly greater. The same principle will hold also where more than one group of linked genes is segregating. Thus, for example, with four genes falling into two groups, each comprising two genes with $d_a = 3$ and $d_b = 1$, and $p = 0.05$ in both cases, the two groups being unlinked with each other, we should find $\hat{k} = 2 \times 0.87 = 1.74$ and $\bar{d} = 2.30$ when both groups were in the dispersed arrangement, and $\hat{k} = 2 \times 1.04 = 2.08$ and $\bar{d} = 3.85$ when both were in the associated arrangement.

So, if we cross two parental lines differing at a number of loci which fall into linked groups, and with the alleles at the loci within the groups preponderantly in the dispersion arrangement, and select for high and low expressions of the character in the descendants of the cross, we expect to pick up and fix recombinants within the groups and so to have replaced the preponderantly dispersed arrangements of the parental groups by preponderantly associated arrangements in the selected lines. Then on estimating k from the cross between the selected lines we would expect to find $\hat{k}$ much the same as that found from the cross of the parent lines themselves, but with $\bar{d}$ increased to an extent corresponding to the effectiveness of the selection in raising and lowering the expression of the character in the high and low selective lines respectively.

This is well illustrated by an experiment described by Mather (1941) in which two lines of *Drosophila melanogaster* were crossed. Beginning with the F_2, selection was practised over thirteen generations for an increased number and over twelve generations for a decreased number of abdominal chaetae. The selected lines were then crossed with each other and an F_2 raised.

The results of this experiment are summarized in Table 59, where the means and variances shown are the averages of males and females. The mean numbers of abdominal chaeta ($\overline{P}_1$ and $\overline{P}_2$) are shown for the two lines that were crossed together for both the original cross, with which

TABLE 59.

$\hat{k}$ and $\overline{d}$ in the original lines and the selection lines derived from their cross, in a selection experiment for abdominal chaetae in *Drosophila melanogaster* (Mather, 1941)

Cross	$\overline{P}_1$	$\overline{P}_2$	$[d]$	V_E	V_{1F2}	D	$\hat{k}$	$\overline{d}$
Original lines	42.24	39.77	1.235	6.412	6.932	1.040	1.5	0.84
Selected lines	46.12	32.85	6.635	7.544	17.469	19.850	2.2	2.99

$$V_E = \tfrac{1}{4}V_{P1} + \tfrac{1}{4}V_{P2} + \tfrac{1}{2}V_{F1}$$

the experiment was started, and the cross between the two selected lines, high and low, derived from that original cross. In each case the non-heritable component of the V_{1F2}, the variance of the F_2, was estimated by combining the variance of P_1, P_2 and F_1 in the F_2 proportions, thus $V_E = \frac{1}{4}V_{P1} + \frac{1}{4}V_{P2} + \frac{1}{2}V_{F1}$. $V_{1F2} - V_E$ is taken as an estimate of $\frac{1}{2}D$. This assumes that H is 0 and so almost certainly overestimates D, but the overestimation is unlikely to be serious since there was little evidence of dominance in these crosses and in any case H makes only half the contribution of D to V_{1F2}. Nevertheless to the extent that H exceeded 0, k will be an underestimate, although the bias will be equal for the two crosses unless the dominance ratio H/D differs between them. It should be noted too, that the estimate of D and hence that of $\hat{k}$ will be less precise in the case of the original cross since the small difference between V_{1F2} and V_E, from which D is found, will render it subject to sampling variation proportionately much greater than in the cross between the selected lines where the difference between V_{1F2} and V_E is much larger.

Despite these necessary provisos, however, the results are clear and striking. In both crosses there are some two, or if we allow for the lowering of the estimate arising from inequality of their effects, perhaps three effective factors, but in the cross of the selected lines the average effect of the factors is about $3\frac{1}{2}$ times as great as in the original cross. The effect of selection has been to build up greatly the effects of the units of inheritance that we can detect and whose number we can estimate by biometrical methods.

These findings have a simple interpretation in terms of linked groups of genes, and indeed as we have seen are to be expected on that basis. They afford us the prime clue to our understanding of how selection acts by rearranging linked combinations of the genes - polygenic combinations as they are called. They also emphasize to us the distinction between the effective factors that we can detect and the genes that we postulate and of which the factors are made up. Effective factors are not genes which can change only by the process (or combination of processes) that we term mutation. Their physical basis lies in the pieces of chromosomes marked and delimited by the genes - all members of the same polygenic system - through whose effects they are recognized. And being pieces of chromosome, they can change their genic content and hence their effects by recombination. They thus have a quality of lability and hence of transcience much greater than that of their constituent genes, which can change only by mutation. True they will be changed by the mutations of their constituent genes, but this is a rarer event than is the recombination which takes place within them as many experiments have shown. Recombination within effective factors rather than mutation of their constituent genes is the basis for understanding the reassortment of polygenic variability and hence of response to selection. It is a basis, too, which allows us to understand the way in which selection appears to create the polygenic variability upon which response to its impact depends (Mather, 1973) and this is reflected in the combinations of constancy, or near constancy, of $\hat{k}$ with change in $\bar{d}$.

Furthermore, since the basis of the effective factor is a piece of chromosome, we must expect it to include not only a number of linked genes which are members of the same polygenic system and hence affecting the expression of the character through which the factor is recognized, but also other genes, members of other polygenic systems affecting other characters. The properties in action of an effective factor can thus transcend the properties of the individual genes of which it is composed, in at least two ways. First a factor comprising two or more genes in a preponderantly dispersed arrangement, each of which is dominant in the same direction, can show overdominance as a factor even though none of the individual genes shows overdominance. This is indeed one of the classical explanations of the occurrence of heterosis in the F_1 of two inbred lines and of course by the same token also of inbreeding depression. Secondly, taking into account the admixture of different polygenic combinations in the same piece of chromosome, the effective factor can show pleiotropy in its action even though none of its constituent genes shows pleiotropic action as an individual. Such a 'pleiotropy' provides a basis for understanding the correlated responses to selection that are so commonly and so extensively observed. But being a pleiotropy that depends on linkage, it can be resolved by recombination, thus those correlated expressions of two or more characters which we recognize as correlated responses to selection can be, and indeed in experiment regularly have been, resolved by giving time and opportunity for recombination to reassort the genic content of the effective factor (Mather, 1973).

38. Other sources of estimates

The estimate of the number of effective factors that we have been discussing (K_I as Mather and Jinks term it) is but one of a number of estimates that can be derived, given an appropriate body of data. One such further estimate can be obtained from the dominance properties of the genes. Where $[h]$ is the deviation of the F_1 mean from the mid-parent we can find $\hat{k} = [h]^2/H$, in just the same way as we have found $\hat{k} = [d]^2/D$. This further estimate has, however, no advantage over the one we have been using: its properties and limitations are essentially the same except that it will not be affected by the association or dispersion of like alleles between the parents, which is resolvable by recombination, but by the reinforcement or opposition of the dominance of the different genes in the system which is not similarly resolvable. In this sense it is of less use than $\hat{k} = [d]^2/D$ and its inferiority is all the greater because whatever the uncertainty arising from the sampling variation of $[d]$ and D, that of $\hat{k} = [h]^2/H$ will be greater since the sampling variations of $[h]$ and especially H will generally be greater than those of their counterparts.

Of more interest are the estimates of k arrived at in quite a different way. If, for example, we have available the variances of a number of F_3 families raised from different individuals of F_2, we can estimate k as

$$\hat{k} = \frac{{}_H V_{2F3}^2}{{}_H V_{VF3}}$$

where V_{2F3} is of course the mean variance of these F_3's. V_{VF3} is the variance of the variances and the subscript H denotes that it is the heritable portion of the variances about which we are talking (M and J, p. 311). This estimate is the K_2 of Mather and Jinks. Similar estimates can be derived from the variances of groups of S_3 and also second back-cross families. This type of estimate has one great advantage over the estimates we have been using: in the absence of linkage it is unaffected by the association or dispersion of alleles in the parental lines, just as in the absence of linkage D is unaffected by association or dispersion although $[d]$ is. It has, however, two disadvantages over and above its requirement for an F_3 or similar generation to be raised. The first is that it is more affected by inequality of the effects of the genes than is the $\hat{k}$ we have been using. This is, however, probably not actually so serious a matter as the fact that to obtain it we have to estimate not just V_{2F3} and V_{VF3}, but the heritable components of these variances $_HV_{2F3}$ and $_HV_{VF3}$. To do so involves the use of a number of corrections based on the estimates of non-heritable variation obtained from parents and F_1, and these corrections may not be small by comparison with the F_3 variances that they are used to correct. The estimate of k that is ultimately obtained is thus likely to be subject to a proportionately greater standard error and the confidence with which it can be used is correspondingly reduced.

Useful estimates of this kind can nevertheless be obtained where the necessary data are available (M and J, pp. 319–24), and if obtainable they can be put to very good use because as we have already noted they are not affected by the dispersion of like alleles between the parents. Now going back to the estimate of k that we have chiefly been discussing in this chapter, we found $\hat{k} = [d]^2/D$, which can be rewritten as $[d]^2 = \hat{k}D$. Given, therefore that we have an independent estimate of k and knowing D, we can find $[d]^2$ and hence $[d]$. And given further that the estimate of k we are using is independent of the association or dispersion of the genes the $[d]$ that we do find will in fact be an estimate of $S(d)$. So if we cross two parent lines and, by raising from them F_2, back-crosses and F_3's or any other combination of families that will give us the value of D together with a $\hat{k}$ of the second kind (K_2) we can calculate $S(d)$. This will tell us whether we can expect to produce lines that will transcend the parent lines in their expression of the character we are considering, and indeed how far they will so transcend them. The value of such information to a breeder concerned to enhance or diminish the expression of the character needs no emphasis.

Still a third basic method of estimating the number of effective factors has recently been developed by Jinks and Towey (1976). It depends on ascertaining the proportion of individuals in a generation, say the F_2, which are heterozygous for at least one gene - or rather one effective factor. This proportion is found by raising a progeny, an F_3 family for example, from each of a number of individuals in the F_2. Two individuals are selfed from each F_3 family, and if the two F_4's so produced differ in either mean or variance (or of course both) in respect of the character under observation, the two F_3 individuals must have had different genotypes and the F_2 individual which gave rise to the F_3 from which they were taken must have been heterozygous for at least one effective factor. Thus the proportion of F_2 heterozygous for at least one unit is ascertained and assuming no linkage of the effective factors their number can be estimated. Once again, the estimate must be minimal since there could have been gene differences too small to detect by families of the size used; but equally the estimate will be unaffected by dispersion of the genic differences in the parents. It can then be used in the same way as the K_2 estimates derived from the variances of F_3 or similar families and there are fewer corrections to be made in the process of estimation, although of course it requires continuing the experiment for an extra generation to F_4.

9

Conclusion

39. Designing the experiments

In the foregoing chapters we have seen how additive gene effects, dominance, non-allelic interaction, linkage and $g \times e$ interaction may be represented in biometrical terms, how they may be distinguished both from one another and from non-heritable effects and how they may be detected and measured biometrically in the descendants of single-crosses and in randomly breeding populations. We have not covered the full range of genetical phenomena - we have, for example, not touched on sex-linkage, cytoplasmic inheritance and maternal effects, haploidy and polyploid inheritance. But we have seen enough of biometrical genetics to appreciate that it is capable of dealing with any of the many phenomena that genetic analysis has taught us to recognize: we proceed by introducing the appropriate parameters into the specifications of the phenotypic expression of the character and then, by comparing the appropriate statistics from relevant types of families, go on to test and measure these parameters.

In the case of haploid inheritance, the biometrical analysis is actually simpler than with diploids, since dominance and all dominance related interactions no longer enter into the specification: we can dispense with h, j, l, g_h and all the other parameters representing dominance and dominance based effects. In other cases like those of sex-linkage and cytoplasmic inheritance more parameters are needed; but this need not complicate the experiments unduly for although we require more statistics from which to construct the additional equations of estimation made necessary by these additional parameters, we do not need additional types of family since we can obtain the extra statistics by subdividing the observations according to sex within the families and generations in the case of sex-linkage or according to the direction of the initial cross in the case of cytoplasmic effects. In still other cases, however, the complexity of the experiments and analysis is greatly increased by the introduction of the further parameters into the specification. More, and perhaps many more, types of family may be needed to provide the necessary statistics. We can see this without even going beyond the phenomena we have discussed in the earlier chapters, for if we wish to examine the capacity for digenic interaction between linked loci to account for the behaviour of a character we need some 20 different types of family of appropriate kinds to carry out the test. This indeed requires a complex experiment and a complex analysis; but it has been done (Jinks and Perkins, 1969), and so in its own way it serves to emphasize the point that in principle any genetical phenomena can be accommodated in the biometrical approach, albeit at a price.

This prospective price serves in its turn to emphasize various points about experimental design and analytical procedure. Thus, sex-linkage and cytoplasmic effects can be detected by appropriate comparisons between reciprocals from crossing two true-breeding lines. It behoves us therefore to raise and compare reciprocal F_1's and where the individuals are unisexual to record the sexes separately and compare them in these F_1's, in order to ascertain whether any complexities arising from these phenomena must be taken into account in planning later generations of the experiment. In other cases observations on certain specific combinations of relationship are needed if the analysis is to be complete and we must ensure that these appear in our data. Thus, in randomly breeding populations the covariance of parent and offspring is $\frac{1}{4}D_R$ and that of half-sibs is $\frac{1}{8}D_R$ while full-sibs give a covariance of $\frac{1}{4}D_R + \frac{1}{16}H_R$. The comparison of either the parent/offspring or the half-sib covariance with that of full-sibs can give evidence of dominance, but that between the parent/offspring and half-sib covariances cannot do so. We must therefore ensure that data on full-sibs are obtained, whether we include both of the other relations or only one of them. To take a second example, non-allelic interaction can be detected from the means of parents, F_1, F_2 and back-crosses. But the detection of linkage requires not merely the use of variances from segregating generations, but variances of at least two ranks. In the absence of interaction the most informative comparison is of the heritable portion of V_{1F2} with the heritable portion of V_{2F3}, since in the absence of linkage ${}_HV_{1F2} = 2\,{}_HV_{2F3}$. If they differ significantly, linkage must be judged to be operative and the sign of the difference will tell us its preponderant phase. So the experiment should be designed to facilitate this comparison being made with maximum efficiency; and the further comparison of ${}_HV_{1F2}$ with ${}_HV_{1F3}$ (which must be available if ${}_HV_{2F3}$ can be found) will provide an additional test of whether there are detectable differences between second degree statistics of the same rank but from different generations, such as would result from non-allelic interaction.

Other examples could readily be given of the need for care in the genetical design of the experiments, that is for designing them so as to permit and facilitate the detection and measurement of the genetical phenomena at issue. We must also, however, pay attention to the statistical design, that is to the adoption of a design which will provide a valid estimate of error variation against which the genetically important comparisons can be tested, and which will enable us as far as possible to make these comparisons with the maximum efficiency permitted by the numbers of individuals and families that available resources permit us to raise and observe. The provision of a valid estimate of error will always entail a design which allows a valid estimate of the non-heritable component of variation, which is of course error variation for the purpose of genetical analysis, and this may in its turn put restrictions on the way we raise, for example, plants from the time the seed is sown (see M and J, pp. 338-9).

Non-heritable variation is not, however, the only type of error variation to be taken into account: the effects of genetical phenomena which the experiment was not designed to take into account and which may indeed not have been recognized as operative in the material in question, may also be affecting the comparison, the testing of which is the prime purpose of the experiment. There is thus a need to obtain more than one set of comparisons which will reflect the phenomena under investigation and to compare these with one another to see whether the phenomena are adequate to account for the heritable differences observed, or whether the sets are sufficiently different from one another to require us to recognize that further unspecified genetical complications exist. This genetico-statistical point is well illustrated by the joint scaling tests that we discussed in Section 9. There we were testing the additive-dominance model, with a view first to detecting and measuring additive and dominance components represented explicitly by $[d]$ and $[h]$ in the formulations, and secondly to testing whether these, taken together with the non-heritable variation, were adequate to account for the differences observed among the mean measurements of parents, F_1, F_2 and back-crosses. The comparison between means of the parents, $\bar{P}_1 - \bar{P}_2$ would itself have been sufficient to establish that $[d]$ was significant and that additive variation was therefore present, just as $F_1 - \frac{1}{2}(\bar{P}_1 + \bar{P}_2)$ by showing that $[h]$ was significant would have established that dominance was operative. The introduction of F_2, B_1 and B_2 in principle allowed further, independent comparisons from which [d] and [h] could be measured and compared among themselves and with the estimates from P_1, P_2 and F_1 for consistency. This was implicitly done by the $\chi^2_{[3]}$ for goodness of fit (Table 6) which tests whether there are detectable sources of genetical variation, and hence genetical phenomena beyond additive gene effects and dominance, displaying their effects in these data, that is whether there is genetical as well as non-heritable error variation. It is thus a test of the adequacy of the genetical formulation, for which purpose we in fact used it.

The example we discussed yielded no evidence of such further genetical complication: the additive-dominance formulation was adequate. But had it proved to be inadequate we could have gone on to use the degrees of freedom on which the test of adequacy was based for the introduction into the formulation of further parameters specifying additional genetical effects, which could then have been measured and tested for their adequacy to account for the residual variation, as indeed we did in the later example of Section 20.

Turning to the precision of the statistics we obtain from our families and of the genetically meaningful comparisons that we seek to make among them, it is obvious that, other things being equal, the bigger the experiment the greater the precision that will be obtained. But resources are not infinite and those available, whether of land, labour, cultural or analytical facilities, will always set a limit to the size of the experiment we can carry out and hence to the precision of the results and the information we can obtain. In this connection, therefore, our task is basically that of designing the experiment so that the maximum of relevant information is obtained from the number of individuals that we can raise, observe and analyse. Having decided on the types of family that must be included to provide the statistics and comparisons needed to answer the genetical questions we have in mind, and to provide the estimates of error variation, heritable and non-heritable, that our tests of significance will require, we must next decide how we shall apportion the individuals between numbers of families of each of the various kinds and numbers of individuals within each of these families.

Taking a simple example, if genetical considerations require us to use F_3 families in order to estimate V_{1F3} and V_{2F3}, we can obtain estimates of these two variances with approximately equal precision by raising $n/2$ families each of two individuals where n individuals can be raised in all. For some purposes this would be the thing to do, but for others it would not: to take another case, if we merely needed an estimate of $\bar{F}_3$ or we were concerned solely with separating the D, H and E components of variation, we might decide that n families each of one individual would be preferable. Here, however, the matter of biological manipulation enters again for it is as easy with, say, *Drosophila* to raise an S_3 family of 40 or 50 as it is to raise a family of one whereas every additional family means an additional mating, and an additional culture. With self-pollinating plants on the other hand little labour is involved in producing F_3 seed and n single plant families are not much more troublesome to raise than a single family of n plants. We might observe a further restraint also imposed by the biology of the species. S_3 is the nearest to an F_3 generation that can be obtained from *Drosophila* or any other dioecious species, whereas the crossing needed to produce an S_3 may be very troublesome in naturally self-pollinating species of plants like wheat, barley or tomatoes, in which F_3's are easy to obtain. Thus many considerations enter into designing experiments in biometrical genetics to derive the maximum information for the resources available. Sometimes we can use earlier experience to help us, but in general little attention has yet been paid to problems of experimental design in biometrical genetics: some of its problems have been investigated but much remains to be done (M and J, Section 58).

One further point remains to be made about statistical precision. Some of the analyses we have discussed have been of means, and others of the second degree statistics, variances and covariances. Now means are subject to much lower error variances than are second degree statistics and so yield estimates and comparisons of greater precision for any given number of individuals observed. Thus information arising from the analysis of means is easier to obtain, and in that sense more rewarding, than information from the analysis of second degree statistics, and for this reason the value in biometrical genetics of anything beyond the analysis of means has on occasion been denied. This would indeed be a fair point if first degree and second degree statistics were merely alternative ways of obtaining the same genetical information, but we have in fact seen that they are not. Means provide us for example with an estimate of $[d]$, which may range anywhere from 0 to $S(d)$ according to the distribution of like alleles between the parents, and an estimate of $[h]$ which will be reduced by any opposition in the direction of dominance between genes at different loci. We can never, therefore, be confident of obtaining a measure of average dominance from the analysis of means. Second degree statistics on the other hand yield estimates of D and H, which in the absence of linkage, are unaffected either by the distribution of alleles between the parents or by differences in the direction of dominance at different loci. So $\sqrt{(H/D)}$ is in principle always able to provide a measure of the average dominance. Furthermore, linkage can be detected and measured only by using second degree statistics, and the analysis of randomly breeding populations too can be achieved only by the use of second degree statistics. So, far from being no more than alternative sources of the same information, first and second degree statistics provide different and complementary information. To deny the value of one because it is statistically more troublesome is merely to shut one's eyes to this complementary quality. If we are to gain the genetical information we require we must be prepared to face the statistical problems it entails and seek to overcome them.

40. Concepts and uses

Biometrical genetics requires statistically valid analyses of results from experiments designed to this end. It also of course requires that the analyses are genetically meaningful, and this in its turn makes demands on the design of the experiments, as we saw in the previous section. The genetical requirement goes deeper, however: ultimately it must imply that the genetical formulations of the means, variances and covariances that we observe and compare in the analysis, must be derived from the basic principles of genetics and be expressed in terms of parameters that properly represent and quantify acceptable genetical phenomena.

These basic genetical principles and (at any rate in the main) the genetical phenomena that we might seek to incorporate were obtained not from the biometrical study of continuous variation, but by using the Mendelian approach of observing the properties and inter-relations of individually recognizable and hence individually traceable genes. Indeed this must be so, for biometrical genetics could not of itself have laid the wide genetical foundation on which the biometrical analyses rest. True, the concept of equilinear transmission from male and female parents could have been established by biometrical means, and particulate inheritance could have been inferred from the excess of variation in F_2 over that shown by inbred parents and their F_1. It would, however, have been virtually impossible to establish with any confidence the precise rules of segregation of these particles or the variety of their relations to one another in hereditary transmission. Neither could the chromosome theory have been established as we know it, nor the nature of linkage and the mechanism of recombination understood. Dominance and at least some interactions could have been demonstrated, but a precise basis for their quantitative analysis would still have been lacking.

Conceptually therefore biometrical genetics is the child of Mendelian genetics. But it is the partner, too, since the concepts can seldom be taken over and used just as they are. Ambiguities must first be removed from them, and they must be refined and adapted to yield the parameters by which they are represented and quantified for biometrical use. To take an example, in the early days of genetics dominance was the capacity of a gene to over-ride the expression of its recessive allele in a heterozygote, whose phenotype was thus the same as that of the homozygote for the dominant gene. It was soon recognized that the heterozygote might be intermediate in phenotype between the two homozygotes, and this was termed incomplete dominance, as distinct from the, by implication, customary complete dominance; but no attempt was made to recognize degrees of incompleteness or to define the absence of dominance, the possibility of which is clearly implied as a special case of incompleteness. Later, the *Drosophila* geneticists came to use the term in yet a different way, any mutant gene which displayed its presence by changing the phenotype when heterozygous with the wild-type allele being described as dominant, without any reference to the relation the phenotype of the heterozygote might bear to that of the mutant homozygote. This new usage had the curious result of a mutant gene whose expression was not always readily recognizable in heterozygotes being sometimes described as 'dominant, but better used as a recessive'. Clearly, although the concept of dominance obviously had to be brought into biometrical genetics, it equally obviously had to be given a consistent and quantitatively precise definition before it could be so used. We have seen in earlier chapters how this is done in the form of the ratio h/d, and this leads us to recognize the fundamental distinction between the phenotypic relation an F_1 bears to its parents as expressed by $[h]/[d]$ (the potence ratio as it has been called) and the dominance ratios, h/d, of the gene-pairs which contribute to that relation. It also emphasizes a further feature of any ratio which depends on the relations between three or more measurements, namely the general dependence of the ratio on the choice of the scale used in making the measurements. We saw the consequence of this dependence for the dominance ratio in the example on p. 46, where the choice of scale could affect its magnitude and even change its sign.

If we take non-allelic interaction as a second illustration, a further point is brought out. Various kinds of digenic interaction were early recognized by geneticists through the aberrations they produced in the classical 9:3:3:1 ratio expected in F_2, and indeed the interpretation of these aberrant ratios in terms of interaction made a major contribution to the establishment of Mendelian inheritance as both ubiquitous and virtually exclusive. As recognized by the early geneticists, these ratios involved not only complete dominance but also complete dependence in expression of specific alleles at the two loci, and the interactions were classified as being of one kind or another according to the particular combination of dominance and dependence that the various genes displayed. Thus before it could be used in biometrical genetics, not only had the concept to be extended to allow of partial as well as complete interaction, but a framework also had to be found which could accomodate all the types of interaction, complementary, duplicate and so on, and so avoid the need to treat each of them separately from the rest. This framework is provided by the recognition of three basic types of interaction: $d \times d$, or i-type; $d \times h$ or j-type; and $h \times h$ or l-type, in the way we saw in Section 19. All the classical types of interaction are definable in terms of i, j and l, each of which makes its own characteristic contributions to the means and variances of the different generations and families, thus affording not merely the means of specifying and quantifying incomplete interaction but also of combining into a single formulation the different types and degrees of interaction that might be expressed by the member genes of a polygenic system when taken two at a time. Again we can see the distinction between, on the one hand, the gross or overall interaction properties taking together all the genes by which two parent lines differ, as represented by $[i]$, $[j]$ and $[l]$, and on the other hand the interaction properties of individual pairs of genes, as represented by the individual i's, j's and l's. Furthermore the representation of interactions is now readily extensible to trigenic or even higher orders, should this be required.

Since the biometrical and Mendelian techniques seek to analyse genetical situations in terms of the same principles and the same phenomena, it has sometimes been assumed that the biometrical approach is no more than a rival alternative to the Mendelian. In principle it is true that some situations, normally and properly dealt with by classical means, could be handled biometrically. This would be the case, for example, where a single gene difference in some readily measurable character, like stature, was involved. But to use the biometrical approach rather than the Mendelian where the genetical classes are easily recognized by inspection would, to say the least, be inefficient and even tortuous; and in the absence of some compelling reason, Mendelian analysis would always be preferred to biometrical in such a case. Indeed to see the two approaches as rival alternatives is to miss the point that each technique of analysis has its own field of application, to which the other is less suited or even impossible to adapt. We should recall that biometrical genetics began and has been developed for the genetical analysis of continuous variation. Even this can be expedited (although seldom if ever carried through to completion) where appropriate special means of genetical or cytological manipulation are available, as we saw when discussing the information to be gained from direct assays of variation in the sternopleural chaeta number of *Drosophila melanogaster* (Section 3). Such analyses of continuous variation require, however, special marker genes and chromosome types which are available in only a few well investigated species of animals and plants, and only to a limited extent in most even of these. They require, too, elaborate and lengthy breeding programmes which are justifiable only for special reasons such as obtained, for example, in the experiments to which we have just referred, where we were concerned to ascertain the detailed nature of the genetical control of the variation and the distribution of the controlling elements between and, as far as possible, within the chromosome.

Thus in all but special cases in a few species, the genetical investigation and understanding of continuous variation must require the use of methods that only biometrical genetics provides. Without these methods continuous variation can be neither probed nor manipulated efficiently. Furthermore, since biometrical genetics neither depends on nor makes use of the recognition, through their effects, of individual gene differences, its analyses will cover all the variation shown by a character, whether non-heritable or heritable, stemming from genes of large effect or small or for that matter from transmissable agents which are not nuclear genes. The completeness of this coverage must often mean complexity in the analyses themselves and in the experiments upon whose results the analyses are based. It is however, this capacity for not only dealing with continuous variation but for clarifying, measuring, analysing and understanding the totality of the variation shown by a character which gives to biometrical genetics its place in our armoury of genetical methodologies for investigating the properties and changes of variability, its adjustment in the wild and its manipulation in those species that we have brought into domestication.

Glossary of symbols and abbreviations

A-a (B-b, etc.)	A pair of alleles, a gene pair, a single gene difference. A is the allele which increases and a that which decreases the expression of the character.
A (B, C etc.)	Individual scaling tests.
b	Regression coefficient.
c	A measure of gene association in the parental lines of a diallel.
d	The departure of one of a pair of corresponding homozygotes from their mid-point or mid-parent (m). It is positive for the homozygote carrying the increasing allele and negative for that carrying the decreasing allele. The relevant gene pair may be denoted by a subscript: thus AA departs from m by d_a and aa departs from m by $-d_a$.
$[d]$	The departure of one of a pair of true breeding parental lines from their mid-parent (m). The parent with the greater expression departs by $[d]$ and that with the lower expression by $-[d]$. $[d]$ is the sum, taking sign into account, of the d's of all the relevant genes carried by the larger parent.
D	$= S(d^2)$ The genetically additive component of variation.
D_P	$= S(4uvd^2)$ The genetical component of variation among the parents of a diallel. $D_P = D$ when all $u = v = \frac{1}{2}$.

D_R	$= S(4uv[d + (v - u)h]^2)$. The statistically additive component in a randomly breeding population. $D_R = D$ when all $u = v = \frac{1}{2}$
D_W	$= S(4uvd[d + (v - u)h])$. The genetical component in $\bar{W}_r$ from a diallel. $D_W = D$ when all $u = v = \frac{1}{2}$.
df	Degrees of freedom.
e	The departure from m ascribable to the effect of the environment, averaged over all genotypes. A biological measure of the environment.
e_X	The departure from m of genotype X ascribable to the environment.
E	The non-heritable component of variation. E_w is the non-heritable component ascribable to differences expressed within a family and E_b that is ascribable to differences as expressed between families. $E_1 = E_w$, $E_2 = E_b + \frac{1}{n} E_w$.
F	$= S(dh)$.
g	The departure from m ascribable to interaction of the genotype and environment. g_d is the interaction of d or $[d]$ with e. g_h is the interaction of h or $[h]$ with e.
G	$= S(g^2)$. The component of variation ascribable to genotype X environment interaction. $G_d = S(g_d^2)$. $G_h = S(g_h^2)$.
h	The departure of the heterozygote from the mid-parent, m. h takes sign, h_a for example being positive when in its expression of the character Aa is nearer AA than to aa, and negative when it is nearer to aa than to AA.
$[h]$	The departure of an F_1 from the mid-parent of the true breeding lines of which it may be regarded as a cross. $[h]$ is the sum taking sign into account of the h's of all the relevant genes.
H	$= S(h^2)$ The dominance component of variation.
H_R	$= S(16u^2v^2h^2)$ in a randomly breeding population. $H_R = H$ where all $u = v = \frac{1}{2}$.

i_{ab} The departure from m ascribable to the hom × hom interaction of A-a and B-b. The interaction of d_a and d_b.

I (i) $= S(i^2)$. The component of variation ascribable to hom × hom interaction.
(ii) The sum of the interaction terms in V_{1F2} etc.

I_R The corresponding component in a randomly breeding population. $I_R = I$ when all $u = v = \frac{1}{2}$.

j_{ab} (j_{ba}) The departures from m ascribable to hom × het interaction of A-a and B-b. j_{ab} is the interaction of d_a and h_b, j_{ba} that of d_b and h_a.

J $= S(j^2)$. The component of variation ascribable to hom × het interaction.

J_R The corresponding component in a randomly breeding population. $J_R = J$ when all $u = v = \frac{1}{2}$.

k The number of gene pairs in which two true breeding lines differ. $\hat{k}$, the estimate of k, is the number of effective factors.

l_{ab} The departure from m ascribable to the het × het interaction of A-a and B-b. The interaction of h_a and h_b.

L $= S(l^2)$. The component of variation ascribable to het × het interaction.

L_R The corresponding component in a randomly breeding population. $L_R = L$ when all $u = v = \frac{1}{2}$.

m The mid-point between the expressions of the character in two true-breeding lines. Commonly termed the mid-parent (but see also pp. 102).

MS Mean square.

n The number of individuals in a family. Similarly n' is the number of families in a group.

p (q) A recombination frequency. p_{ab} is the frequency of recombination between A-a and B-b. $1 - p = q$.

P Probability, in relation to a test of significance.

r	(i) Correlation coefficient. (ii) Measure of association of the genes in which two true-breeding lines differ. Where the lines differ by k gene pairs of equal effect, and the larger parent nevertheless carries reducing alleles at k' of them, $r = \frac{1}{k}(k - 2k')$. $r = 1$ for complete association and 0 for maximum dispersion.
S	Indicates summation.
SCP	Sum of cross products.
SS	Sum of squares.
t	The ratio of a quantity to its estimated standard error. A test of significance.
T	Heritability, T_n being the narrow and T_b the broad heritability.
$u\,(v)$	The frequency of the increasing allele. $1 - u = v$. Thus u_a is the frequency of allele A and v_a that of a.
V	A variance, the relevance of which is indicated by a subscript. Thus V_{P1} is the variance of P_1 (the larger parent), V_{F1} that of F_1, V_{1F2} that of F_2 etc.
V_R	The variance of a randomly breeding population, with V_{SR} the variance of a full sibs, V_{HSR} the variance of half-sibs etc.
V_r	The variance of an array in a diallel. $\overline{V}_r$ is the mean variance of all arrays and $V_{\bar{r}}$ the variance of array means.
VR	Variance ratio. A test of significance.
W	A covariance the relevance of which is indicated by a subscript. Thus W_{1F23} is the covariance of an F_2 individual with the mean of the F_3 derived from it, etc.
W_R	A covariance in a randomly breeding population, the relevance of which is indicated by a subscript. Thus W_{POR} is the covariance of parent and offspring, W_{SR} that of full sibs, W_{HSR} that of half-sibs etc.

W_r — The covariance with the non-recurrent parent given by an array in a diallel. $\overline{W}_r$ is the mean covariance of all arrays.

α — A measure of the variation in magnitude of a set of d's.

θ — A measure of the intensity of complementary and duplicate type interactions between gene pairs. Where $d_a = d_b = h_a = h_b = d$, $i_{ab} = j_{ab} = j_{ba} = l_{ab} = \theta d$, θ is positive for complementary and negative for duplicate type interaction.

The mean of a family or generation is denoted by a bar over the designation of that family or generation. The mean of parent P_1 is $\overline{P}_1$, of F_1 is $\overline{F}_1$, of F_2 is $\overline{F}_2$ etc.

References

ÅKERMAN, Å. (1922). Untersuchungen über eine in direktem Sonnenlichte nicht lebensfahige Sippe von *Avena sativa. Hereditas* **3**, 147–77.

BATESON, W. (1909). *Mendel's Principles of Heredity.* University Press, Cambridge.

CALIGARI, P.D.S. and MATHER, K. (1975). Genotype-environment interaction: III. Interactions in *Drosophila melanogaster. Proc. R. Soc. Lond. B.* **191**, 387–411.

CAVALLI, L.L. (1952). An analysis of linkage in quantitative inheritance. *Quantitative Inheritance* (Ed. E.C.R. Reeve and C.H. Waddington) pp. 135–44. HMSO, London.

DARLINGTON, C.D. and MATHER, K. (1949). *The Elements of Genetics.* Allen and Unwin, London.

DAVIES, R.W. (1971). The genetic relationship of two quantitative characters in *Drosophila melanogaster.* II. Location of the effects. *Genetics* **69**, 363–75.

EAVES, L.J. (1975). Testing models for variation in intelligence. *Heredity* **34**, 132–6.

EAVES, L.J., LAST, K., MARTIN, N.G. and JINKS, J.L. (1977). A Progressive approach to non-additivity and genotype-environmental covariance in the analysis of human differences. *Br. J. Mathematical and Statistical Psychology* (in press).

FALCONER, D.S. (1960). *Introduction to Quantitative Genetics.* Oliver and Boyd, Edinburgh.

FISHER, R.A. (1918). The correlations between relatives on the supposition of Mendelian inheritance. *Trans. R. Soc. Edinb.* **52**, 399–433.

FISHER, R.A. (1946). *Statistical Methods for Research Workers* (10th Edn). Oliver and Boyd, Edinburgh.

GALTON, F. (1889). *Natural Inheritance.* Macmillan, London.

GOLDSCHMIDT, R. (1938). *Physiological Genetics.* McGraw-Hill, New York and London.

GRÜNEBERG, H. (1952). Genetical studies on the skeleton of the mouse; IV. Quasi-continuous variations. *J. Genet.* **51**, 95–114.

HAYMAN, B.I. (1960). Maximum likelihood estimation of genetic components of variation. *Biometrics* **16**, 369–81.

HOGBEN, L. (1933). *Nature and Nurture.* Allen and Unwin, London.

HOLT, S.B. (1952). Genetics of dermal ridges: Inheritance of total finger ridge-count. *Ann. Eugen.* **17**, 140–61.

JINKS, J.L. and CONNOLLY, V. (1973). Selection for specific and general response to environmental differences. *Heredity* **30**, 33–40.

JINKS, J.L. and FULKER, D.W. (1970). A comparison of the biometrical genetical, MAVA and classical approaches to the analysis of human behaviour. *Psychol. Bull.* **73**, 311–49.

JINKS, J.L. and PERKINS, J.M. (1969). The detection of linked epistatic genes for a metrical trait. *Heredity* **24**, 465–75.

JINKS, J.L., PERKINS, J.M. and BREESE, E.L. (1969). A general method of detecting additive, dominance and epistatic variation for metrical traits: II. Application to inbred lines. *Heredity* **24**, 45–57.

JINKS, J.L. and TOWEY, P. (1976). Estimating the number of genes in a polygenic system by genotype assay. *Heredity* **37**, 69–81.

JOHANNSEN, W. (1909). *Elemente der exakten Erblichkeitslehre.* Fischer, Jena.

KEARSEY, M.J. (1965). Biometrical analysis of a random mating population: A comparison of five experimental designs. *Heredity* **20**, 205–35.

KEARSEY, M.J. and JINKS, J.L. (1968). A general method of detecting additive, dominance and epistatic variation for metrical traits: I. Theory. *Heredity* **23**, 403–9.

LAW, C.N. (1967). The location of genetic factors controlling a number of quantitative characters in wheat. *Genetics* **56**, 445–61.

MARTIN, N.G. (1975). The inheritance of scholastic abilities in a sample of twins. *Ann. hum. Genet.* **39**, 219–29.

MATHER, K. (1941). Variation and selection of polygenic characters. *J. Genet.* **41**, 159–93.

MATHER, K. (1949). *Biometrical Genetics* (1st Edn.) Methuen, London.

MATHER, K. (1967). *The Elements of Biometry.* Methuen, London.

MATHER, K. (1973). *Genetical Structure of Populations.* Chapman and Hall, London.

MATHER, K. (1974). Non-allelic interaction in continuous variation of randomly breeding populations. *Heredity* **32**, 414–19.

MATHER, K. and HARRISON, B.J. (1949). The manifold effect of selection. *Heredity* **3**, 1–52 and 131–62.

MATHER, K. and JINKS, J.L. (1971). *Biometrical Genetics* (2nd Edn.) Chapman and Hall, London. (This reference is abbreviated to M and J in the text.)

PERKINS, J.M. and JINKS, J.L. (1970). Detecting and estimation of genotype-environmental, linkage and epistatic components of variation for a metrical trait. *Heredity* **25**, 157–77.

POWERS, L. (1951). Gene analysis by the partitioning method when interactions of genes are involved. *Bot. Gaz.* **113**, 1–23.

ROBINSON, H.F., COMSTOCK, R.E. and HARVEY, P.H. (1949). Estimates of heritability and the degree of dominance in corn. *Agron. J.* **41**, 353–9.

SEARLE, S.R. (1966). *Matrix Algebra for Biologists.* Wiley, New York.

SHIELDS, J. (1962). *Monozygotic Twins.* Oxford, University Press.

SPICKETT, S.G. (1963). Genetic and developmental studies of a quantitative character. *Nature* **199**, 870–3.

STURTEVANT, A.H. (1925). The effects of unequal crossing over at the bar locus in *Drosophila. Genetics* **10**, 117–47.

THODAY, J.M. (1961). Location of polygenes. *Nature* **191**, 368–70.

VAN DER VEEN, J.H. (1959). Tests of non-allelic interaction and linkage for quantitative characters in generations derived from two diploid pure lines. *Genetica* **30**, 201–32.

WOLSTENHOLME, D.R. and THODAY, J.M. (1963). Effects of disruptive selection: VII. A third chromosome polymorphism. *Heredity* **18**, 413–31.

Index